Math Made #1
Nice-n-Easy Books™

In This Book:
- ## Number Systems
- ## Sets
- ## Integers
- ## Fractions
- ## Decimals

"MATH MADE NICE-n-EASY #1" is one in a series of books designed to make the learning of math interesting and fun. For help with additional math topics, see the complete series of "MATH MADE NICE-n-EASY" titles.

Based on U.S. Government
Teaching Materials

Research & Education Association
61 Ethel Road West
Piscataway, New Jersey 08854

MATH MADE NICE-N-EASY BOOKS™
BOOK #1

Year 2003 Printing

Printed in the United States of America

Library of Congress Control Number 99-70142

International Standard Book Number 0-87891-200-2

MATH MADE NICE-N-EASY is a trademark of
Research & Education Association, Piscataway, New Jersey 08854

WHAT "MATH MADE NICE-N-EASY" WILL DO FOR YOU

The "Math Made Nice-n-Easy" series simplifies the learning and use of math and lets you see that math is actually interesting and fun. This series of books is for people who have found math scary, but who nevertheless need some understanding of math without having to deal with the complexities found in most math textbooks.

The "Math Made Nice-n-Easy" series of books is useful for students and everyone who needs to acquire a basic understanding of one or more math topics. For this purpose, the series is divided into a number of books which deal with math in an easy-to-follow sequence beginning with basic arithmetic, and extending through pre-algebra, algebra, and calculus. Each topic is described in a way that makes learning and understanding easy.

Almost everyone needs to know at least some math at work, or in a course of study.

For example, almost all college entrance tests and professional exams require solving math problems. Also, almost all occupations (waiters, sales clerks, office people) and all crafts (carpentry, plumbing, electrical) require some ability in math problem solving.

The "Math Made Nice-n-Easy" series helps the reader grasp quickly the fundamentals that are needed in using math. The reader is led by the hand, step-by-step, through the various concepts and how they are used.

By acquiring the ability to use math, the reader is encouraged to further his/her skills and to forget about any initial math fears.

The "Math Made Nice-n-Easy" series includes material originated by U.S. Government research and educational efforts. The research was aimed at devising tutoring and teaching methods for educating government personnel lacking a technical and/or mathematical background. Thanks for these efforts are due to the U.S. Bureau of Naval Personnel Training.

Dr. Max Fogiel
Program Director

Contents

Chapter 3

SIGNED NUMBERS 54

Chapter 4

COMMON FRACTIONS 79

CHAPTER 1

NUMBER SYSTEMS AND SETS

Many of us have areas in our mathematics background that are hazy, barely understood, or troublesome. Thus, while it may at first seem beneath your dignity to read chapters on fundamental arithmetic, these basic concepts may be just the spots where your difficulties lie. These chapters attempt to treat the subject on an adult level that will be interesting and informative.

COUNTING

Counting is such a basic and natural process that we rarely stop to think about it. The process is based on the idea of ONE-TO-ONE CORRESPONDENCE, which is easily demonstrated by using the fingers. When children count on their fingers, they are placing each finger in one-to-one correspondence with one of the objects being counted. Having outgrown finger counting, we use numerals.

NUMERALS

Numerals are number symbols. One of the

simplest numeral systems is the Roman numeral system, in which tally marks are used to represent the objects being counted. Roman numerals appear to be a refinement of the tally method still in use today. By this method, one makes short vertical marks until a total of four is reached; when the fifth tally is counted, a diagonal mark is drawn through the first four marks. Grouping by fives in this way is reminiscent of the Roman numeral system, in which the multiples of five are represented by special symbols.

A number may have many "names." For example, the number 6 may be indicated by any of the following symbols: 9 - 3, 12/2, 5 + 1, or 2 x 3. The important thing to remember is that a number is an idea; various symbols used to indicate a number are merely different ways of expressing the same idea.

POSITIVE WHOLE NUMBERS

The numbers which are used for counting in our number system are sometimes called natural numbers. They are the positive whole numbers, or to use the more precise mathematical term, positive INTEGERS. The Arabic numerals from 0 through 9 are called digits, and an integer may have any number of digits. For example, 5, 32, and 7,049 are all integers. The number of digits in an integer indicates its rank; that is, whether it is "in the hundreds," "in the thousands," etc. The idea of ranking numbers in terms of tens, hundreds, thousands, etc., is based on the PLACE VALUE concept.

PLACE VALUE

Although a system such as the Roman numeral system is adequate for recording the results of counting, it is too cumbersome for purposes of calculation. Before arithmetic could develop as we know it today, the following two important concepts were needed as additions to the counting process:

1. The idea of 0 as a number.
2. Positional notation (place value).

Positional notation is a form of coding in which the value of each digit of a number depends upon its position in relation to the other digits of the number. The convention used in our number system is that each digit has a higher place value than those digits to the right of it.

The place value which corresponds to a given position in a number is determined by the BASE of the number system. The base which is most commonly used is ten, and the system with ten as a base is called the decimal system (decem is the Latin word for ten). Any number is assumed to be a base-ten number, unless some other base is indicated. One exception to this rule occurs when the subject of an entire discussion is some base other than ten. For example, in the discussion of binary (base two) numbers later in this chapter, all numbers are assumed to be binary numbers unless some other base is indicated.

DECIMAL SYSTEM

In the decimal system, each digit position in a number has ten times the value of the position adjacent to it on the right. For example, in the number 11, the 1 on the left is said to be in the "tens place," and its value is 10 times as great as that of the 1 on the right. The 1 on the right is said to be in the "units place," with the understanding that the term "unit" in our system refers to the numeral 1. Thus the number 11 is actually a coded symbol which means "one ten plus one unit." Since ten plus one is eleven, the symbol 11 represents the number eleven.

Figure 1-1 shows the names of several digit positions in the decimal system. If we apply this nomenclature to the digits of the integer 235, then this number symbol means "two hundreds plus three tens plus five units." This number may be expressed in mathematical symbols as follows:

$$2 \times 10 \times 10 + 3 \times 10 \times 1 + 5 \times 1$$

Notice that this bears out our earlier statement: each digit position has 10 times the value of the position adjacent to it on the right.

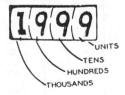

Figure 1-1.—Names of digit positions.

The integer 4,372 is a number symbol whose meaning is "four thousands plus three hundreds plus seven tens plus two units." Expressed in mathematical symbols, this number is as follows:

$$4 \times 1000 + 3 \times 100 + 7 \times 10 + 2 \times 1$$

This presentation may be broken down further, in order to show that each digit position as 10 times the place value of the position on its right, as follows:

$$4 \times 10 \times 100 + 3 \times 10 \times 10 + 7 \times 10 \times 1 + 2 \times 1$$

The comma which appears in a number symbol such as 4,372 is used for "pointing off" the digits into groups of three beginning at the right-hand side. The first group of three digits on the right is the units group; the second group is the thousands group; the third group is the millions group; etc. Some of these groups are shown in table 1-1.

Table 1-1.—Place values and grouping.

Billions group	Millions group	Thousands group	Units group
Hundred billions Ten billions Billions	Hundred millions Ten millions Millions	Hundred thousands Ten thousands Thousands	Hundreds Tens Units

By reference to table 1-1, we can verify that 5,432,786 is read as follows: five million, four hundred thirty-two thousand, seven hundred eighty-six. Notice that the word "and" is not necessary when reading numbers of this kind.

Practice problems:

1. Write the number symbol for seven thousand two hundred eighty-one.
2. Write the meaning, in words, of the symbol 23,469.
3. If a number is in the millions, it must have at least how many digits?
4. If a number has 10 digits, to what number group (thousands, millions, etc.) does it belong?

Answers:

1. 7,281
2. Twenty-three thousand, four hundred sixty-nine.
3. 7
4. Billions

BINARY SYSTEM

The binary number system is constructed in the same manner as the decimal system. However, since the base in this system is two, only two digit symbols are needed for writing numbers. These two digits are 1 and 0. In order to understand why only two digit symbols are needed in the binary system, we may make some observations about the decimal system and then generalize from these.

One of the most striking observations about number systems which utilize the concept of place value is that there is no single-digit symbol for the base. For example, in the decimal system the symbol for ten, the base, is 10. This symbol is compounded from two digit symbols, and its meaning may be interpreted as "one base plus no units." Notice the implication of this where other bases are concerned: Every system uses the same symbol for the base, namely 10. Furthermore, the symbol 10 is not called "ten" except in the decimal system.

Suppose that a number system were constructed with five as a base. Then the only digit symbols needed would be 0, 1, 2, 3, and 4. No single-digit symbol for five is needed, since the symbol 10 in a base-five system with place value means "one five plus no units." In general, in a number system using base N, the largest number for which a single-digit symbol is needed is N minus 1. Therefore, when the base is two the only digit symbols needed are 1 and 0.

An example of a binary number is the symbol 101. We can discover the meaning of this symbol by relating it to the decimal system. Figure 1-2 shows that the place value of each digit position in the binary system is two times the place value of the position adjacent to it on the right. Compare this with figure 1-1, in which the base is ten rather than two.

Placing the digits of the number 101 in their respective blocks on figure 1-2, we find that

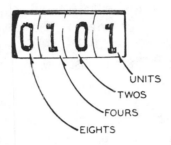

UNITS
TWOS
FOURS
EIGHTS

Figure 1-2.—Digit positions
in the binary system.

101 means "one four plus no twos plus one unit."
Thus 101 is the binary equivalent of decimal 5.
If we wish to convert a decimal number, such
as 7, to its binary equivalent, we must break it
into parts which are multiples of 2. Since 7 is
equal to 4 plus 2 plus 1, we say that it "contains" one 4, one 2, and one unit. Therefore
the binary symbol for decimal 7 is 111.

The most common use of the binary number
system is in electronic digital computers. All
data fed to a typical electronic digital computer
is converted to binary form and the computer
performs its calculations using binary arithmetic rather than decimal arithmetic. One of
the reasons for this is the fact that electrical
and electronic equipment utilizes many switch-
ing circuits in which there are only two operating conditions. Either the circuit is "on" or it
is "off," and a two-digit number system is
ideally suited for symbolizing such a situation.

8

Practice problems:
1. Write the decimal equivalents of the binary numbers 1101, 1010, 1001, and 1111.
2. Write the binary equivalents of the decimal numbers 12, 7, 14, and 3.

Answers:
1. 13, 10, 9, and 15
2. 1100, 111, 1110, and 11

SETS

Any serious study of mathematics leads the student to investigate more than one text and more than one way of approaching each new topic. At the time of printing of this course, much emphasis is being placed on so-called modern math in the public schools. Consequently, the trainee who uses this course is likely to find considerable material, in his parallel reading, which uses the ideas and terminology of the "new" math.

In the following paragraphs, a very brief introduction to some of the set theory of modern math is presented. Although the remainder of this course is not based on set theory, this brief introduction should help in making the transition from traditional methods to newer, experimental methods.

DEFINITIONS AND SYMBOLS

The word "set" implies a collection or grouping of similar objects or symbols. The objects in a set have at least one characteristic in common, such as similarity of appearance or purpose. A set of tools would be an example of a group of objects not necessarily similar in appearance but similar in purpose. The objects or symbols in a set are called members or ELEMENTS of the set.

The elements of a mathematical set are usually symbols, such as numerals, lines, or points. For example, the positive integers greater than zero and less than 5 form a set, as follows:

$$\{1, 2, 3, 4\}$$

Notice that braces are used to indicate sets. This is often done where the elements of the set are not too numerous.

Since the elements of the set $\{2, 4, 6\}$ are the same as the elements of $\{4, 2, 6\}$, these two sets are said to be equal. In other words, equality between sets has nothing to do with the order in which the elements are arranged. Furthermore, repeated elements are not necessary. That is, the elements of $\{2, 2, 3, 4\}$ are simply 2, 3, and 4. Therefore the sets $\{2, 3, 4\}$ and $\{2, 2, 3, 4\}$ are equal.

Practice problems:
1. Use the correct symbols to designate the set of odd positive integers greater than 0 and less than 10.
2. Use the correct symbols to designate the set of names of days of the week which do not

10

contain the letter "s".

3. List the elements of the set of natural numbers greater than 15 and less than 20.

4. Suppose that we have sets as follows:

$$A = \{1, 2, 3\} \qquad C = \{1, 2, 3, 4\}$$

$$B = \{1, 2, 2, 3\} \qquad D = \{1, 1, 2, 3\}$$

Which of these sets are equal?

Answers:
1. $\{1, 3, 5, 7, 9\}$
2. $\{Monday, Friday\}$
3. 16, 17, 18, and 19
4. $A = B = D$

SUBSETS

Since it is inconvenient to enumerate all of the elements of a set each time the set is mentioned, sets are often designated by a letter. For example, we could let S represent the set of all positive integers greater than 0 and less than 10. In symbols, this relationship could be stated as follows:

$$S = \{1, 2, 3, 4, 5, 6, 7, 8, 9\}$$

Now suppose that we have another set, T, which comprises all positive even integers less than 10. This set is then defined as follows:

$$T = \{2, 4, 6, 8\}$$

Notice that every element of T is also an ele-

ment of S. This establishes the SUBSET relationship; T is said to be a subset of S.

POSITIVE INTEGERS

The most fundamental set of numbers is the
set of positive integers. This set comprises
the counting numbers (natural numbers) and includes, as subsets, all of the sets of numbers
which we have discussed. The set of natural
numbers has an outstanding characteristic: it
is infinite. This means that the successive
elements of the set continue to increase in size
without limit, each number being larger by 1
than the number preceding it. Therefore there
is no "largest" number; any number that we
might choose as larger than all others could be
increased to a larger number simply by adding
1 to it.

One way to represent the set of natural numbers symbolically would be as follows:

$$\{1, 2, 3, 4, 5, 6, \ldots\}$$

The three dots, called ellipsis, indicate that the
pattern established by the numbers shown continues without limit. In other words, the next
number in the set is understood to be 7, the
next after that is 8, etc.

POINTS AND LINES

In addition to the many sets which can be
formed with number symbols, we frequently

find it necessary in mathematics to work with sets composed of points or lines.

A point is an idea, rather than a tangible object, just as a number is. The mark which is made on a piece of paper is merely a symbol representing the point. In strict mathematical terms, a point has no dimensions (physical size) at all. Thus a pencil dot is only a rough picture of a point, useful for indicating the location of the point but certainly not to be confused with the ideal.

Now suppose that a large number of points are placed side by side to form a "string." Picturing this arrangement by drawing dots on paper, we would have a "dotted line." If more dots were placed between the dots already in the string, with the number of dots increasing until we could not see between them, we would have a rough picture of a line. Once again, it is important to emphasize that the picture is only a symbol which represents an ideal line. The ideal line would have length but no width or thickness.

The foregoing discussion leads to the conclusion that a line is actually a set of points. The number of elements in the set is infinite, since the line extends in both directions without limit.

The idea of arranging points together to form a line may be extended to the formation of planes (flat surfaces). A mathematical plane may be considered as the result of placing an infinite number of straight lines side by side, with no space between the lines. Thus the plane is a set of lines. Another way of defining a plane in terms of sets is to consider the plane as the result of placing points side by side in all directions. In this case, the plane is a set of points and the points comprising any line in the plane form a subset.

Line Segments and Rays

When we draw a "line," label its end points A and B, and call it "line AB," we really mean LINE SEGMENT AB. A line segment is a subset of the set of points comprising a line.

When a line is considered to have a starting point but no stopping point (that is, it extends without limit in one direction), it is called a RAY. A ray is not a line segment, because it does not terminate at both ends; it may be appropriate to refer to a ray as a "half-line."

As in the case of a line segment, a ray is a subset of the set of points comprising a line. All three—lines, line segments, and rays—are subsets of the set of points comprising a plane.

THE NUMBER LINE

Among the many devices used for representing a set of numbers, one of the most useful is the number line. To illustrate the construction of a number line, let us place the elements of the set of natural numbers in one-to-one correspondence with points on a line. Since the natural numbers are equally spaced, we select points such that the distances between them are equal. The starting point is labeled 0, the next point is labeled 1, the next 2, etc., using the natural numbers in normal counting order. (See fig. 1-3.) Such an arrangement is often referred to as a scale, a familiar example being the scale on a thermometer.

Thus far in our discussion, we have not mentioned any numbers other than integers. The number line is an ideal device for picturing the

Figure 1-3.—A number line.

relationship between integers and other numbers such as fractions and decimals. It is clear that many points, other than those representing integers, exist on the number line. Examples are the points representing the numbers 1/2 (located halfway between 0 and 1) and 2.5 (located halfway between 2 and 3).

An interesting question arises, concerning the "in-between" points on the number line: How many points (numbers) exist between any two integers? To answer this question, suppose that we first locate the point halfway between 0 and 1, which corresponds to the number 1/2. Then let us locate the point halfway between 0 and 1/2, which corresponds to the number 1/4. The result of the next such halving operation would be 1/8, the next 1/16, etc. If we need more space to continue our halving operations

15

on the number line, we can enlarge our "picture" and then continue.

It soon becomes apparent that the halving process could continue indefinitely; that is, without limit. In other words, the number of points between 0 and 1 is infinite. The same is true of any other interval on the number line. Thus, between any two integers there is an infinite set of numbers other than integers. If this seems physically impossible, considering that even the sharpest pencil point has some width, remember that we are working with ideal points, which have no physical dimensions whatsoever.

Although it is beyond the scope of this course to discuss such topics as orders of infinity, it is interesting to note that the set of integers contains many subsets which are themselves infinite. Not only are the many subsets of numbers other than integers infinite, but also such subsets as the set of all odd integers and the set of all even integers. By intuition we see that these two subsets are infinite, as follows: If we select a particular odd or even integer which we think is the largest possible, a larger one can be formed immediately by merely adding 2.

Perhaps the most practical use for the number line is in explaining the meaning of negative numbers. Negative numbers are discussed in detail in chapter 3 of this course.

CHAPTER 2

POSITIVE INTEGERS

The purpose of this chapter is to review the methods of combining integers. We have already used one combination process in our discussion of counting. We will extend the idea of counting, which is nothing more than simple addition, to develop a systematic method for adding numbers of any size. We will also learn the meaning of subtraction, multiplication, and division.

ADDITION AND SUBTRACTION

In the following discussion, it is assumed that the reader knows the basic addition and subtraction tables, which present such facts as the following: $2 + 3 = 5$, $9 + 8 = 17$, $8 - 3 = 5$, etc.

The operation of addition is indicated by a plus sign (+) as in $8 + 4 = 12$. The numbers 8 and 4 are ADDENDS and the answer (12) is their SUM. The operation of subtraction is indicated by a minus sign (-) as in $9 - 3 = 6$. The number 9 is the MINUEND, 3 is the SUBTRAHEND, and the answer (6) is their DIFFERENCE.

REGROUPING

Addition may be performed with the addends arranged horizontally, if they are small enough and not too numerous. However, the most common method of arranging the addends is to place them in vertical columns. In this arrangement, the units digits of all the addends are alined vertically, as are the tens digits, the hundreds digits, etc. The following example shows three addends arranged properly for addition:

$$357$$
$$1,845$$
$$\underline{\quad 22}$$

It is customary to draw a line below the last addend, placing the answer below this line. Subtraction problems are arranged in columns in the same manner as for addition, with a line at the bottom and the answer below this line.

Carry and Borrow

Problems involving several addends, with two or more digits each, usually produce sums in one or more of the columns which are greater than 9. For example, suppose that we perform the following addition:

$$357$$
$$845$$
$$\underline{\quad 22}$$
$$1,224$$

18

The answer was found by a process called "carrying." In this process extra digits, generated when a column sum exceeds 9, are carried to the next column to the left and treated as addends in that column. Carrying may be explained by grouping the original addends. For example, 357 actually means 3 hundreds plus 5 tens plus 7 units. Rewriting the problem with each addend grouped in terms of units, tens, etc., we would have the following:

$$
\begin{array}{r}
300 + \ 50 + \ 7 \\
800 + \ 40 + \ 5 \\
20 + \ 2 \\
\hline
1{,}100 + 110 + 14
\end{array}
$$

The "extra" digit in the units column of the answer represents 1 ten. We regroup the columns of the answer so that the units column has no digits representing tens, the tens column has no digits representing hundreds, etc., as follows:

$$
\begin{aligned}
1{,}100 + 110 + 14 &= 1{,}100 + 110 + 10 + 4 \\
&= 1{,}100 + 120 + 4 \\
&= 1{,}100 + 100 + 20 + 4 \\
&= 1{,}200 + 20 + 4 \\
&= 1{,}000 + 200 + 20 + 4 \\
&= 1{,}224
\end{aligned}
$$

When we carry the 10 from the expression $10 + 4$ to the tens column and place it with the 110 to make 120, the result is the same as if we had added 1 to the digits 5, 4, and 2 in the tens column of the original problem. Therefore, the thought process in addition is as follows: Add the 7, 5, and 2 in the units column, getting a sum of 14. Write down the 4 in the units column of the answer and carry the 1 to the tens column. Mentally add the 1 along with the other digits in the tens column, getting a sum of 12. Write down the 2 in the tens column of the answer and carry the 1 to the hundreds column. Mentally add the 1 along with the other digits in the hundreds column, getting a sum of 12. Write down the 2 in the hundreds column of the answer and carry the 1 to the thousands column. If there were other digits in the thousands column to which the 1 could be added, the process would continue as before. Since there are no digits in the thousands column of the original problem, this final 1 is not added to anything, but is simply written in the thousands place in the answer.

The borrow process is the reverse of carrying and is used in subtraction. Borrowing is not necessary in such problems as 46 - 5 and 58 - 53. In the first problem, the thought process may be "5 from 6 is 1 and bring down the 4 to get the difference, 41." In the second problem, the thought process is "3 from 8 is 5" and "5 from 5 is zero," and the answer is 5. More explicitly, the subtraction process in these examples is as follows:

$$40 + 6 \qquad\qquad 50 + 8$$
$$\underline{5} \qquad\qquad \underline{50 + 3}$$
$$40 + 1 = 41 \qquad 0 + 5 = 5$$

This illustrates that we are subtracting units from units and tens from tens.

Now consider the following problem where borrowing is involved:

$$\begin{array}{r} 43 \\ \underline{8} \end{array}$$

If the student uses the borrowing method, he may think "8 from 13 is 5 and bring down 3 to get the difference, 35." In this case what actually was done is as follows:

$$\begin{array}{r} 30 + 13 \\ \underline{8} \end{array}$$

$$30 + 5 = 35$$

A 10 has been borrowed from the tens column and combined with the 3 in the units column to make a number large enough for subtraction of the 8. Notice that borrowing to increase the value of the digit in the units column reduces the value of the digit in the tens column by 1.

Sometimes it is necessary to borrow in more than one column. For example, suppose that we wish to subtract 2,345 from 5,234. Grouping the minuend and subtrahend in units, tens, hundreds, etc., we have the following:

$$\begin{array}{r} 5,000 + 200 + 30 + 4 \\ \underline{2,000 + 300 + 40 + 5} \end{array}$$

Borrowing a 10 from the 30 in the tens column,

we regroup as follows:

$$5,000 + 200 + 20 + 14$$
$$2,000 + 300 + 40 + 5$$

The units column is now ready for subtraction. By borrowing from the hundreds column, we can regroup so that subtraction is possible in the tens column, as follows:

$$5,000 + 100 + 120 + 14$$
$$2,000 + 300 + 40 + 5$$

In the final regrouping, we borrow from the thousands column to make subtraction possible in the hundreds column, with the following result:

$$4,000 + 1,100 + 120 + 14$$
$$2,000 + 300 + 40 + 5$$
$$2,000 + 800 + 80 + 9 = 2,889$$

In actual practice, the borrowing and regrouping are done mentally. The numbers are written in the normal manner, as follows:

$$5,234$$
$$-2,345$$
$$2,889$$

The following thought process is used: Borrow from the tens column, making the 4 become 14. Subtracting in the units column, 5 from 14 is 9. In the tens column, we now have a 2 in the minuend as a result of the first borrowing operation. Some students find it helpful at first to cancel any digits that are reduced as a result of borrowing, jotting down the digit of next lower value just above the canceled digit. This has been done in the following example:

$$
\begin{array}{r}
4\ 12 \\
\cancel{5},\cancel{2}\cancel{3}\cancel{4} \\
-2,345 \\
\hline
2,889
\end{array}
$$

After canceling the 3, we proceed with the subtraction, one column at a time. We borrow from the hundreds column to change the 2 that we now have in the tens column into 12. Subtracting in the tens column, 4 from 12 is 8. Proceeding in the same way for the hundreds column, 3 from 11 is 8. Finally, in the thousands column, 2 from 4 is 2.

Practice problems. In problems 1 through 4, add the indicated numbers. In problems 5 through 8, subtract the lower number from the upper.

1. Add 23, 468, 7, and 9,045.

2. 129
    958
    787
    436
    ___

3. 9,497
    6,364
    4,269
    9,785
    ___

4. 67,856
    22,851
    44,238
    97,156
    ___

5. 709
    594
    ___

6. 8,700
    5,008
    ___

7. 7,928
    5,349
    ___

8. 75,168
    28,089
    ___

Answers:

1. 9,543   2. 2,310   3. 29,915   4. 232,101
5. 115   6. 3,692   7. 2,579   8. 47,079

Denominate Numbers

Numbers that have a unit of measure associated with them, such as yard, kilowatt, pound, pint, etc., are called DENOMINATE NUMBERS. The word "denominate" means the numbers have been given a name; they are not just abstract symbols. To add denominate numbers, add all units of the same kind. Simplify the result, if possible. The following example illustrates the addition of 6 ft 8 in. to 4 ft 5 in.:

$$\begin{array}{rr} 6 \text{ ft} & 8 \text{ in.} \\ \underline{4 \text{ ft}} & \underline{5 \text{ in.}} \\ 10 \text{ ft} & 13 \text{ in.} \end{array}$$

Since 13 in. is the equivalent of 1 ft 1 in., we regroup the answer as 11 ft 1 in.

A similar problem would be to add 20 degrees 44 minutes 6 seconds to 13 degrees 22 minutes 5 seconds. This is illustrated as follows:

$$\begin{array}{rrrr} 20 \text{ deg} & 44 \text{ min} & 6 \text{ sec} \\ \underline{13 \text{ deg}} & \underline{22 \text{ min}} & \underline{5 \text{ sec}} \\ 33 \text{ deg} & 66 \text{ min} & 11 \text{ sec} \end{array}$$

This answer is regrouped as 34 deg 6 min 11 sec.

24

Numbers must be expressed in units of the same kind, in order to be combined. For instance, the sum of 6 kilowatts plus 1 watt is not 7 kilowatts nor is it 7 watts. The sum can only be indicated (rather than performing the operation) unless some method is used to write these numbers in units of the same value.

Subtraction of denominate numbers also involves the regrouping idea. If we wish to subtract 16 deg 8 min 2 sec from 28 deg 4 min 3 sec, for example, we would have the following arrangement:

$$\begin{array}{r} 28 \text{ deg } 4 \text{ min } 3 \text{ sec} \\ -16 \text{ deg } 8 \text{ min } 2 \text{ sec} \\ \hline \end{array}$$

In order to subtract 8 min from 4 min we regroup as follows:

$$\begin{array}{r} 27 \text{ deg } 64 \text{ min } 3 \text{ sec} \\ -16 \text{ deg } 8 \text{ min } 2 \text{ sec} \\ \hline 11 \text{ deg } 56 \text{ min } 1 \text{ sec} \end{array}$$

Practice problems. In problems 1 and 2, add. In problems 3 and 4, subtract the lower number from the upper.

1. 6 yd 2 ft 7 in.
        1 ft 9 in.
    2 yd   10 in.

2. 9 hr 47 min 51 sec
    3 hr 36 min 23 sec
    5 hr 15 min 23 sec

3. 15 hr 25 min 10 sec
    -6 hr 50 min 35 sec

4. 125 deg
    47 deg 9 min 14 sec

Answers:

1. 9 yd 2 ft 2 in.
2. 18 hr 39 min 37 sec
3. 8 hr 34 min 35 sec
4. 77 deg 50 min 46 sec

Mental Calculation

Mental regrouping can be used to avoid the necessity of writing down some of the steps, or of rewriting in columns, when groups of one-digit or two-digit numbers are to be added or subtracted.

One of the most common devices for rapid addition is recognition of groups of digits whose sum is 10. For example, in the following problem two "ten groups" have been marked with braces:

$$
\begin{array}{r}
7 \\
\left.\begin{array}{r} 6 \\ 4 \end{array}\right\} \ 10 \\
5 \\
\left.\begin{array}{r} 1 \\ 9 \end{array}\right\} \ 10 \\
\hline
\end{array}
$$

To add this column as grouped, you would say to yourself, "7, 17, 22, 32." The thought should be just the successive totals as shown above and not such cumbersome steps as "7 + 10, 17, + 5, 22, + 10, 32."

When successive digits appear in a column and their sum is less than 10, it is often convenient to think of them, too, as a sum rather than separately. Thus, if adding a column in which the sum of two successive digits is 10 or less, group them as follows:

$$\left.\begin{array}{c} 3 \\ 1 \\ 1 \end{array}\right\} 5$$

$$\left.\begin{array}{c} 8 \\ 1 \end{array}\right\} 9$$

$$\left.\begin{array}{c} 4 \\ 6 \end{array}\right\} 10$$

The thought process here might be, as shown by the grouping, "5, 14, 24."

Practice problems. Add the following columns from the top down, as in the preceding example:

1.	2	2.	4	3.	88	4.	57
	7		6		36		32
	3		7		59		64
	6		8		82		97
	4		1		28		79
	1		8		57		44

Answers, showing successive mental steps:

1. 2, 12, 22, 23 - - Final answer, 23
2. 10, 17, 26, 34 - - Final answer, 34

3. Units column: 14, 23, 33, 40 - - Write down 0, carry 4.

Tens column: 12, 20, 30, 35 - - Final answer, 350.

4. Units column: 9, 20, 29, 33 - - Write down 3, carry 3.

Tens column: 8, 17, 26, 37 - - Final answer, 373.

SUBTRACTION.—In an example such as 73 - 46, the conventional approach is to place 46 under 73 and subtract units from units and tens from tens, and write only the difference without the intermediate steps. To do this, the best method is to begin at the left. Thus, in the example 73 - 46, we take 40 from 73 and then take 6 from the result. This is done mentally, however, and the thought would be "73, 33, 27," or "33, 27." In the example 84 - 21 the thought is "64, 63" and in the example 64 - 39 the thought is "34, 25."

Practice problems. Mentally subtract and write only the difference:

1. 47 - 24
2. 69 - 38
3. 87 - 58
4. 86 - 73
5. 82 - 41
6. 30 - 12

Answers, showing successive mental steps:

1. 27, 23 - - Final answer, 23
2. 39, 31 - - Final answer, 31
3. 37, 29 - - Final answer, 29
4. 16, 13 - - Final answer, 13
5. 42, 41 - - Final answer, 41
6. 20, 18 - - Final answer, 18

MULTIPLICATION AND DIVISION

Multiplication may be indicated by a multiplication sign (x) between two numbers, a dot between two numbers, or parentheses around one or both of the numbers to be multiplied. The following examples illustrate these methods:

$$6 \times 8 = 48$$
$$6 \cdot 8 = 48$$
$$6(8) = 48$$
$$(6)(8) = 48$$

Notice that when a dot is used to indicate multiplication, it is distinguished from a decimal point or a period by being placed above the line of writing, as in example 2, whereas a period or decimal point appears on the line. Notice also that when parentheses are used to indicate multiplication, the numbers to be multiplied are spaced closer together than they are when the dot or x is used.

In each of the four examples just given, 6 is the MULTIPLIER and 8 is the MULTIPLICAND. Both the 6 and the 8 are FACTORS, and the more modern texts refer to them this way. The "answer" in a multiplication problem is the PRODUCT; in the examples just given, the product is 48.

Division usually is indicated either by a division sign (÷) or by placing one number over another number with a line between the numbers, as in the following examples:

$$1. \ 8 \div 4 = 2$$

$$2. \ \frac{8}{4} = 2$$

The number 8 is the DIVIDEND, 4 is the DIVI-SOR, and 2 is the QUOTIENT.

MULTIPLICATION METHODS

The multiplication of whole numbers may be thought of as a short process of adding equal numbers. For example, 6(5) and 6 x 5 are read as six 5 s. Of course we could write 5 six times and add, but if we learn that the result is 30 we can save time. Although the concept of adding equal numbers is quite adequate in explaining multiplication of whole numbers, it is only a special case of a more general definition, which will be explained later in multiplication involving fractions.

Grouping

Let us examine the process involved in multiplying 6 times 27 to get the product 162. We first arrange the factors in the following manner:

$$\begin{array}{r} 27 \\ \underline{\times 6} \\ 162 \end{array}$$

The thought process is as follows:

1. 6 times 7 is 42. Write down the 2 and carry the 4.

30

2. 6 times 2 is 12. Add the 4 that was carried over from step 1 and write the result, 16, beside the 2 that was written in step 1.

3. The final answer is 162.

Table 2-1 shows that the factors were grouped in units, tens, etc. The multiplication was done in three steps: Six times 7 units is 42 units (or 4 tens and 2 units) and six times 2 tens is 12 tens (or 1 hundred and 2 tens). Then the tens were added and the product was written as 162.

Table 2-1.—Multiplying by a one-digit number.

	Hundreds	Tens	Units
6(27) = 162		2	7
			6
		4	2
	1	2	
	1	6	2

In preparing numbers for multiplication as in table 2-1, it is important to place the digits of the factors in the proper columns; that is, units must be placed in the units column, tens in tens column, and hundreds in hundreds column. Notice that it is not necessary to write the zero in the case of 12 tens (120) since the 1 and 2 are written in the proper columns. In practice, the addition is done mentally, and just the product is written without the intervening steps.

Multiplying a number with more than two digits by a one-digit number, as shown in table 2-2, involves no new ideas. Three times 6 units is 18 units (1 ten and 8 units), 3 times 0 tens is 0, and 3 times 4 hundreds is 12 hundreds (1

Table 2-2.—Multiplying a three-digit number by a one-digit number.

	Thousands	Hundreds	Tens	Units
3(406) = 1,218		4	0	6
				3
			1	8
	1	2		
	1	2	1	8

thousand and 2 hundreds). Notice that it is not necessary to write the 0 s resulting from the step "3 times 0 tens is 0." The two terminal 0's of the number 1,200 are also omitted, since the 1 and the 2 are placed in their correct columns by the position of the 4.

Partial Products

In the example, 6(8) = 48, notice that the multiplying could be done another way to get the correct product as follows:

$$6(3 + 5) = 6 \times 3 + 6 \times 5$$

That is, we can break 8 into 3 and 5, multiply each of these by the other factor, and add the partial products. This idea is employed in multiplying by a two-digit number. Consider the following example:

$$
\begin{array}{r}
43 \\
\times 27 \\
\hline
1,161
\end{array}
$$

Breaking the 27 into 20 + 7, we have 7 units times 43 plus 2 tens times 43, as follows:

$$43(20 + 7) = (43)(7) + (43)(20)$$

Since 7 units times 43 is 301 units, and 2 tens times 43 is 86 tens, we have the following:

$$
\begin{array}{r}
43 \\
\times 27 \\
\hline
301 = 3 \text{ hundreds, 0 tens, 1 unit} \\
86 = 8 \text{ hundreds, 6 tens} \\
\hline
1,161
\end{array}
$$

As long as the partial products are written in the correct columns, we can multiply beginning from either the left or the right of the multiplier. Thus, multiplying from the left, we have

$$
\begin{array}{r}
43 \\
\times 27 \\
\hline
86 \\
301 \\
\hline
1,161
\end{array}
$$

Multiplication by a number having more places involves no new ideas.

End Zeros

The placement of partial products must be kept in mind when multiplying in problems involving end zeros, as in the following example:

$$
\begin{array}{r}
27 \\
\times 40 \\
\hline
1,080
\end{array}
$$

We have 0 units times 27 plus 4 tens times 27, as follows:

$$
\begin{array}{r}
27 \\
\times 40 \\
\hline
0 \\
108 \\
\hline
1,080
\end{array}
$$

The zero in the units place plays an important part in the reading of the final product. End zeros are often called "place holders" since their only function in the problem is to hold the digit positions which they occupy, thus helping to place the other digits in the problem correctly.

The end zero in the foregoing problem can be accounted for very nicely, while at the same time placing the other digits correctly, by means of a shortcut. This consists of offsetting the 40 one place to the right and then simply bringing down the 0, without using it as a multiplier at all. The problem would appear as follows:

$$\begin{array}{r} 27 \\ \times 40 \\ \hline 1,080 \end{array}$$

If the problem involves a multiplier with more than one end 0, the multiplier is offset as many places to the right as there are end 0 s. For example, consider the following multiplication in which the multiplier, 300, has two end 0 s:

$$\begin{array}{r} 220 \\ \times 300 \\ \hline 66,000 \end{array}$$

Notice that there are as many place-holding zeros at the end in the product as there are place-holding zeros in the multiplier and the multiplicand combined.

Placement of Decimal Points

In any whole number in the decimal system, there is understood to be a terminating mark, called a decimal point, at the right-hand end of the number. Although the decimal point is seldom shown except in numbers involving decimal fractions (covered in chapter 5 of this course), its location must be known. The placement of the decimal point is automatically taken care of when the end 0 s are correctly placed.

Practice problems. Multiply in each of the following problems:

1. 287 x 8
2. 67 x 49
3. 940 x 20

4. 807 x 28
5. 694 x 80
6. 9,241 x 7,800

Answers:

1. 2,296
2. 3,283
3. 18,800

4. 22,596
5. 55,520
6. 72,079,800

DIVISION METHODS

Just as multiplication can be considered as repeated addition, division can be considered as repeated subtraction. For example, if we wish to divide 12 by 4 we may subtract 4 from 12 in successive steps and tally the number of times that the subtraction is performed, as follows:

$$
\begin{array}{r}
12 \\
\underline{4} \;* \\
8 \\
\underline{4} \;* \\
4 \\
\underline{4} \;* \\
0
\end{array}
$$

As indicated by the asterisks used as tally marks, 4 has been subtracted 3 times. This result is sometimes described by saying that "4 is contained in 12 three times."

Since successive subtraction is too cumbersome for rapid, concise calculation, methods which treat division as the inverse of multiplication are more useful. Knowledge of the multiplication tables should lead us to an answer for a problem such as $12 \div 4$ immediately, since we know that 3 x 4 is 12. However, a problem such as $84 \div 4$ is not so easy to solve by direct reference to the multiplication table.

One way to divide 84 by 4 is to note that 84 is the same as 80 plus 4. Thus $84 \div 4$ is the same as $80 \div 4$ plus $4 \div 4$. In symbols, this can be indicated as follows:

$$
4\overline{)80 + 4}^{\,20 + 1}
$$

(When this type of division symbol is used, the quotient is written above the vinculum as shown.)

Thus, 84 divided by 4 is 21.

From the foregoing example, it can be seen that the regrouping is useful in division as well as in multiplication. However, the mechanical procedure used in division does not include writing down the regrouped form of the dividend. After becoming familiar with the process, we find that the division can be performed directly, one digit at a time, with the regrouping taking place mentally. The following example illustrates this:

$$
\begin{array}{r}
14 \\
4\overline{)56} \\
4 \\
\hline
16 \\
16 \\
\hline
\end{array}
$$

The thought process is as follows: "4 is contained in 5 once" (write 1 in tens place over the 5); "one times 4 is 4" (write 4 in tens place under 5, take the difference, and bring down 6); and "4 is contained in 16 four times" (write 4 in units place over the 6). After a little practice, many people can do the work shown under the dividend mentally and write only the quotient, if the divisor has only 1 digit.

The divisor is sometimes too large to be contained in the first digit of the dividend. The following example illustrates a problem of this kind:

$$
\begin{array}{r}
36 \\
7\overline{)252} \\
21 \\
\hline
42 \\
42 \\
\hline
\end{array}
$$

Since 2 is not large enough to contain 7, we divide 7 into the number formed by the first two digits, 25. Seven is contained 3 times in 25; we write 3 above the 5 of the dividend. Multiplying, 3 times 7 is 21; we write 21 below the first two digits of the dividend. Subtracting, 25 minus 21 is 4; we write down the 4 and bring down the 2 in the units place of the dividend. We have now formed a new dividend, 42. Seven is contained 6 times in 42; we write 6 above the 2 of the dividend. Multiplying as before, 6 times 7 is 42; we write this product below the dividend 42. Subtracting, we have nothing left and the division is complete.

Estimation

When there are two or more digits in the divisor, it is not always easy to determine the first digit of the quotient. An estimate must be made, and the resulting trial quotient may be too large or too small. For example, if 1,862 is to be divided by 38, we might estimate that 38 is contained 5 times in 186 and the first digit of our trial divisor would be 5. However, multiplication reveals that the product of 5 and 38 is larger than 186. Thus we would change the 5 in our quotient to 4, and the problem would then appear as follows:

$$
\begin{array}{r}
49 \\
38\overline{)1862} \\
\underline{152} \\
342 \\
\underline{342}
\end{array}
$$

On the other hand, suppose that we had esti-
mated that 38 is contained in 186 only 3 times.
We would then have the following:

$$
\begin{array}{r}
3 \\
38\overline{)1862} \\
114 \\
\hline
72
\end{array}
$$

Now, before we make any further moves in the
division process, it should be obvious that some-
thing is wrong. If our new dividend is large
enough to contain the divisor before bringing
down a digit from the original dividend, then the
trial quotient should have been larger. In other
words, our estimate is too small.

Proficiency in estimating trial quotients is
gained through practice and familiarity with
number combinations. For example, after a
little experience we realize that a close estimate can be made in the foregoing problem by
thinking of 38 as "almost 40." It is easy to see
that 40 is contained 4 times in 186, since 4
times 40 is 160. Also, since 5 times 40 is 200,
we are reasonably certain that 5 is too large
for our trial divisor.

Uneven Division

In some division problems such as $7 \div 3$,
there is no other whole number that, when multiplied by the divisor, will give the dividend.
We use the distributive idea to show how divi-

40

sion is done in such a case. For example, $7 \div 3$ could be written as follows:

$$\frac{(6 + 1)}{3} = \frac{6}{3} + \frac{1}{3} = 2\frac{1}{3}$$

Thus, we see that the quotient also carries one unit that is to be divided by 3. It should now be clear that $3\overline{)37} = 3\overline{)30 + 7}$, and that this can be further reduced as follows:

$$\frac{30}{3} + \frac{6}{3} + \frac{1}{3} = 10 + 2 + \frac{1}{3} = 12\frac{1}{3}$$

In elementary arithmetic the part of the dividend that cannot be divided evenly by the divisor is often called a REMAINDER and is placed next to the quotient with the prefix R. Thus, in the foregoing example where the quotient was $12\frac{1}{3}$, the quotient could be written 12 R 1. This method of indicating uneven division is useful in examples such as the following:

Suppose that $13 is available for the purchase of spare parts, and the parts needed cost $3 each. Four parts can be bought with the available money, and $1 will be left over. Since it is not possible to buy 1/3 of a part, expressing the result as 4 R 1 gives a more meaningful answer than 4 1/3.

Placement of Decimal Points

In division, as in multiplication, the placement of the decimal point is important. Determining the location of the decimal point and the number of places in the quotient can be relatively simple if the work is kept in the proper columns. For example, notice the vertical alinement in the following problem:

$$
\begin{array}{r}
311 \\
31\overline{)9,641} \\
9\ 3 \\
\hline
34 \\
31 \\
\hline
31 \\
31 \\
\hline
\end{array}
$$

We notice that the first two places in the dividend are used to obtain the first place in the quotient. Since 3 is in the hundreds column there are two more places in the quotient (tens place and units place). The decimal point in the quotient is understood to be directly above the position of the decimal point in the dividend. In the example shown here, the decimal point is not shown but is understood to be immediately after the second 1.

Checking Accuracy

The accuracy of a division of numbers can be checked by multiplying the quotient by the divisor and adding the remainder, if any. The result should equal the dividend. Consider the following example:

```
         5203
42/218541        Check:    5203
   210                    x  42

    85                    10406
    84                    20812

   141                   218526
   126                   +  15

    15                   218541
```

DENOMINATE NUMBERS

We have learned that denominate numbers
are not difficult to add and subtract, provided
that units, tens, hundreds, etc., are retained in
their respective columns. Multiplication and
division of denominate numbers may also be
performed with comparative ease, by using the
experience gained in addition and subtraction.

Multiplication

In multiplying denominate numbers by integers, no new ideas are needed. If in the problem 3(5 yd 2 ft 6 in.) we remember that we can
multiply each part separately to get the correct
product (as in the example, 6(8) = 6(3) + 6(5)),
we can easily find the product, as follows:

```
        5 yd  2 ft   6 in.
                x 3
       15 yd  6 ft  18 in.
```

Simplifying, this is

$$17 \text{ yd } 1 \text{ ft } 6 \text{ in.}$$

When one denominate number is multiplied by another, a question arises concerning the products of the units of measurement. The product of one unit times another of the same kind is one square unit. For example, 1 ft times 1 ft is 1 square foot, abbreviated sq ft; 2 in. times 3 in. is 6 sq in.; etc. If it becomes necessary to multiply such numbers as 2 yd 1 ft times 6 yd 2 ft, the foot units may be converted to fractions of a yard, as follows:

$$(2 \text{ yd } 1 \text{ ft})(6 \text{ yd } 2 \text{ ft}) = (2 \ 1/3 \text{ yd})(6 \ 2/3 \text{ yd})$$

In order to complete the multiplication, a knowledge of fractions is needed. Fractions are discussed in chapter 4 of this training course.

Division

The division of denominate numbers requires division of the highest units first; and if there is a remainder, conversion to the next lower unit, and repeated division until all units have been divided.

In the example (24 gal 1 qt 1 pt) ÷ 5, we perform the following steps:

Step 1:
$$\begin{array}{r} 4 \text{ gal} \\ 5\overline{)24 \text{ gal}} \\ \underline{20} \\ 4 \text{ gal (left over)} \end{array}$$

Step 2: Convert the 4 gal left over to 16 qt and add to the 1 qt.

Step 3:
$$5\overline{)17\text{ qt}}^{\quad 3\text{ qt}}$$
$$\underline{15}$$
$$2\text{ qt (left over)}$$

Step 4: Convert the 2 qt left over to 4 pt and add to the 1 pt.

Step 5:
$$5\overline{)5\text{ pt}}^{\quad 1\text{ pt}}$$

Therefore, 24 gal 1 qt 1 pt divided by 5 is 4 gal 3 qt 1 pt.

Practice problems. In problems 1 through 4, divide as indicated. In problems 5 through 8, multiply or divide as indicated.

1. 549 ÷ 9

2. 470/63

3. 25/2,300

4. 64/74,816

5. 4 hr 26 min 16 sec
$$\underline{\qquad\qquad\qquad \text{x } 5}$$

6. 3(4 gal 3 qt 1 pt)

7. 67 deg 43 min 12 sec
$$\overline{\qquad\qquad\qquad\qquad}$$
$$2$$

8. 5/63 lb 11 oz

Answers:

1. 61

2. 7 R 29

45

3. 92

4. 1,169

5. 22 hr 11 min 20 sec

6. 14 gal 2 qt 1 pt

7. 33 deg 51 min 36 sec

8. 12 lb 11 4/5 oz

ORDER OF OPERATIONS

When a series of operations involving addition, subtraction, multiplication, or division is indicated, the order in which the operations are performed is important only if division is involved or if the operations are mixed. A series of individual additions, subtractions, or multiplications may be performed in any order. Thus, in

$$4 + 2 + 7 + 5 = 18$$

or

$$100 - 20 - 10 - 3 = 67$$

or

$$4 \times 2 \times 7 \times 5 = 280$$

the numbers may be combined in any order desired. For example, they may be grouped easily to give

$$6 + 12 = 18$$

and

$$97 - 30 = 67$$

and

$$40 \times 7 = 280$$

A series of divisions should be taken in the order written.

Thus,

$$100 \div 10 \div 2 = 10 \div 2 = 5$$

In a series of mixed operations, perform multiplications first, division next, and finally additions and subtractions.

For example
$$100 \div 4 \times 5 = 100 \div 20 = 5$$

and

$$60 - 25 \div 5 = 60 - 5 = 55$$

Now consider
$$60 - 25 \div 5 + 15 - 100 + 4 \times 10$$
$$= 60 - 25 \div 5 + 15 - 100 + 40$$
$$= 60 - 5 + 15 - 100 + 40$$
$$= 115 - 105$$
$$= 10$$

Notice that $25 \div 5$ could be evaluated at the same time that 4×10 is evaluated, since no other multiplication is to be performed in the first part of the problem.

Practice problems. Evaluate each of the following expressions:

1. $9 \div 3 + 2$
2. $18 - 2 \times 5 + 4$
3. $90 \div 2 \div 9$
4. $75 \div 5 \times 3 \div 5$
5. $7 + 1 - 8 \times 4 \div 16$

Answers:

1. 5   4. 1
2. 12   5. 6
3. 5

MULTIPLES AND FACTORS

Any number that is exactly divisible by a given number is a MULTIPLE of the given number. For example, 24 is a multiple of 2, 3, 4, 6, 8, and 12, since it is divisible by each of these numbers. Saying that 24 is a multiple of 3, for instance, is equivalent to saying that 3 multiplied by some whole number will give 24. Any number is a multiple of itself and also of 1.

Any number that is a multiple of 2 is an EVEN NUMBER. The even numbers begin with 2 and progress by 2 s as follows:

$$2, 4, 6, 8, 10, 12, \ldots$$

Any number that is not a multiple of 2 is an ODD NUMBER. The odd numbers begin with 1 and progress by 2 s, as follows:

$$1, 3, 5, 7, 9, 11, 13, \ldots$$

Any number that can be divided into a given number without a remainder is a FACTOR of the given number. The given number is a multiple of any number that is one of its factors. For example, 2, 3, 4, 6, 8, and 12 are factors of 24. The following four equalities show various combinations of the factors of 24:

$$24 = 24 \cdot 1 \qquad 24 = 8 \cdot 3$$
$$24 = 12 \cdot 2 \qquad 24 = 6 \cdot 4$$

If the number 24 is factored as completely as possible, it assumes the form

$$24 = 2 \cdot 2 \cdot 2 \cdot 3$$

ZERO AS A FACTOR

If any number is multiplied by zero, the product is zero. For example, 5 times zero equals zero and may be written $5(0) = 0$. The zero factor law tells us that, if the product of two or more factors is zero, at least one of the factors must be zero.

PRIME FACTORS

A number that has factors other than itself and 1 is a COMPOSITE NUMBER. For example, the number 15 is composite. It has the factors 5 and 3.

A number that has no factors except itself and 1 is a PRIME NUMBER. Since it is sometimes advantageous to separate a composite

49

number into prime factors, it is helpful to be able to recognize a few prime numbers quickly. The following series shows all the prime numbers up to 60:

1, 2, 3, 5, 7, 11, 13, 17, 19, 23, 29, 31, 37, 41, 43, 47, 53, 59.

Notice that 2 is the only even prime number. All other even numbers are divisible by 2. Notice also that 51, for example, does not appear in the series, since it is a composite number equal to 3 x 17.

If a factor of a number is prime, it is called a PRIME FACTOR. To separate a number into prime factors, begin by taking out the smallest factor. If the number is even, take out all the 2 s first, then try 3 as a factor, etc. Thus, we have the following example:

$$540 = 2 \cdot 270$$
$$= 2 \cdot 2 \cdot 135$$
$$= 2 \cdot 2 \cdot 3 \cdot 45$$
$$= 2 \cdot 2 \cdot 3 \cdot 3 \cdot 15$$
$$= 2 \cdot 2 \cdot 3 \cdot 3 \cdot 3 \cdot 5$$

Since 1 is an understood factor of every number, we do not waste space recording it as one of the factors in a presentation of this kind.

A convenient way of keeping track of the prime factors is in the short division process as follows:

2/540
2/270
3/135
3/45
3/15
5/5
1

If a number is odd, its factors will be odd numbers. To separate an odd number into prime factors, take out the 3 s first, if there are any. Then try 5 as a factor, etc. As an example,

$$5,775 = 3 \cdot 1,925$$
$$= 3 \cdot 5 \cdot 385$$
$$= 3 \cdot 5 \cdot 5 \cdot 77$$
$$= 3 \cdot 5 \cdot 5 \cdot 7 \cdot 11$$

Practice problems:

1. Which of the following are prime numbers and which are composite numbers?

$$25, 7, 18, 29, 51$$

2. What prime numbers are factors of 36?

3. Which of the following are multiples of 3?

$$45, 53, 51, 39, 47$$

4. Find the prime factors of 27.

Answers:

1. Prime: 7, 29
   Composite: 25, 18, 51
2. $36 = 2 \cdot 2 \cdot 3 \cdot 3$
3. 45, 51, 39
4. $27 = 3 \cdot 3 \cdot 3$

Tests for Divisibility

It is often useful to be able to tell by inspection whether a number is exactly divisible by one or more of the digits from 2 through 9. An expression which is frequently used, although it is sometimes misleading, is "evenly divisible." This expression has nothing to do with the concept of even and odd numbers, and it probably should be avoided in favor of the more descriptive expression, "exactly divisible." For the remainder of this discussion, the word "divisible" has the same meaning as "exactly divisible." Several tests for divisibility are listed in the following paragraphs:

1. A number is divisible by 2 if its right-hand digit is even.

2. A number is divisible by 3 if the sum of its digits is divisible by 3. For example, the digits of the number 6,561 add to produce the sum 18. Since 18 is divisible by 3, we know that 6,561 is divisible by 3.

3. A number is divisible by 4 if the number formed by the two right-hand digits is divisible by 4. For example, the two right-hand digits of the number 3,524 form the number 24. Since 24 is divisible by 4, we know that 3,524 is divisible by 4.

4. A number is divisible by 5 if its right-hand digit is 0 or 5.

5. A number is divisible by 6 if it is even and the sum of its digits is divisible by 3. For example, the sum of the digits of 64,236 is 21,

which is divisible by 3. Since 64,236 is also an even number, we know that it is divisible by 6.

6. No short method has been found for determining whether a number is divisible by 7.

7. A number is divisible by 8 if the number formed by the three right-hand digits is divisible by 8. For example, the three right-hand digits of the number 54,272 form the number 272, which is divisible by 8. Therefore, we know that 54,272 is divisible by 8.

8. A number is divisible by 9 if the sum of its digits is divisible by 9. For example, the sum of the digits of 546,372 is 27, which is divisible by 9. Therefore we know that 546,372 is divisible by 9.

Practice problems. Check each of the following numbers for divisibility by all of the digits except 7:

1. 242,431,231,320
2. 844,624,221,840
3. 988,446,662,640
4. 207,634,542,480

Answers: All of these numbers are divisible by 2, 3, 4, 5, 6, 8, and 9.

CHAPTER 3

SIGNED NUMBERS

The positive numbers with which we have worked in previous chapters are not sufficient for every situation which may arise. For example, a negative number results in the operation of subtraction when the subtrahend is larger than the minuend.

NEGATIVE NUMBERS

When the subtrahend happens to be larger than the minuend, this fact is indicated by placing a minus sign in front of the difference, as in the following:

$$12 - 20 = -8$$

The difference, -8, is said to be NEGATIVE. A number preceded by a minus sign is a NEGATIVE NUMBER. The number -8 is read "minus eight." Such a number might arise when we speak of temperature changes. If the temperature was 12 degrees yesterday and dropped 20 degrees today, the reading today would be 12 - 20, or -8 degrees.

Numbers that show either a plus or minus

sign are called SIGNED NUMBERS. An unsigned number is understood to be positive and is treated as though there were a plus sign preceding it.

If it is desired to emphasize the fact that a number is positive, a plus sign is placed in front of the number, as in +5, which is read "plus five." Therefore, either +5 or 5 indicates that the number 5 is positive. If a number is negative, a minus sign must appear in front of it, as in -9.

In dealing with signed numbers it should be emphasized that the plus and minus signs have two separate and distinct functions. They may indicate whether a number is positive or negative, or they may indicate the operation of addition or subtraction.

When operating entirely with positive numbers, it is not necessary to be concerned with this distinction since plus or minus signs indicate only addition or subtraction. However, when negative numbers are also involved in a computation, it is important to distinguish between a sign of operation and the sign of a number.

DIRECTION OF MEASUREMENT

Signed numbers provide a convenient way of indicating opposite directions with a minimum of words. For example, an altitude of 20 ft above sea level could be designated as +20 ft. The same distance below sea level would then be designated as -20 ft. One of the most com-

55

mon devices utilizing signed numbers to indicate direction of measurement is the thermometer.

Thermometer

The Celsius (centigrade) thermometer shown in figure 3-1 illustrates the use of positive and negative numbers to indicate direction of travel above and below 0. The 0 mark is the change-over point, at which the signs of the scale numbers change from - to +.

When the thermometer is heated by the surrounding air or by a hot liquid in which it is placed, the mercury expands and travels up the tube. After the expanding mercury passes 0, the mark at which it comes to rest is read as a positive temperature. If the thermometer is allowed to cool, the mercury contracts. After passing 0 in its downward movement, any mark at which it comes to rest is read as a negative temperature.

Rectangular Coordinate System

As a matter of convenience, mathematicians have agreed to follow certain conventions as to the use of signed numbers in directional measurement. For example, in figure 3-2, a direction to the right along the horizontal line is positive, while the opposite direction (toward the left) is negative. On the vertical line, direction upward is positive, while direction downward is negative. A distance of -3 units along the horizontal line indicates a measurement of 3 units to the left of starting point 0. A distance of -3 units on the vertical line indicates

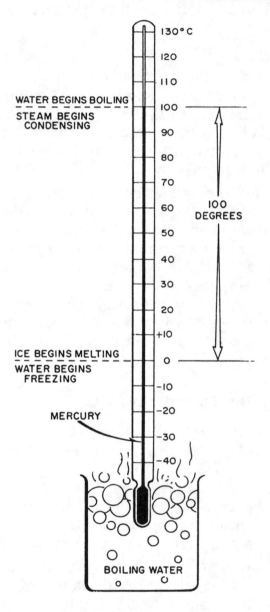

Figure 3-1.—Celsius (centigrade)
temperature scale.

57

a measurement of 3 units below the starting point.

The two lines of the rectangular coordinate system which pass through the 0 position are the vertical axis and horizontal axis. Other vertical and horizontal lines may be included, forming a grid. When such a grid is used for the location of points and lines, the resulting "picture" containing points and lines is called a GRAPH.

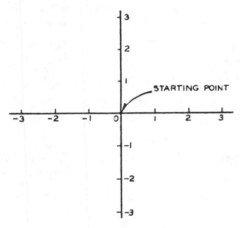

Figure 3-2.—Rectangular
coordinate system.

The Number Line

Sometimes it is important to know the relative greatness (magnitude) of positive and negative numbers. To determine whether a particular number is greater or less than another number, think of all the numbers both positive and negative as being arranged along a horizontal line. (See fig. 3-3.)

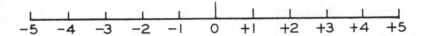

Figure 3-3.—Number line showing both
positive and negative numbers.

Place zero at the middle of the line. Let the positive numbers extend from zero toward the right. Let the negative numbers extend from zero toward the left. With this arrangement, positive and negative numbers are so located that they progress from smaller to larger numbers as we move from left to right along the line. Any number that lies to the right of a given number is greater than the given number. A number that lies to the left of a given number is less than the given number. This arrangement shows that any negative number is smaller than any positive number.

The symbol for "greater than" is >. The symbol for "less than" is <. It is easy to distinguish between these symbols because the symbol used always opens toward the larger number. For example, "7 is greater than 4" can be written 7 > 4 and "-5 is less than -1" can be written -5 < -1.

Absolute Value

The ABSOLUTE VALUE of a number is its numerical value when the sign is dropped. The absolute value of either +5 or -5 is 5. Thus, two numbers that differ only in sign have the same absolute value.

The symbol for absolute value consists of

59

two vertical bars placed one on each side of the number, as in $|-5| = 5$. Consider also the following:

$$|4 - 20| = 16$$
$$|+7| = |-7| = 7$$

The expression $|-7|$ is read "absolute value of minus seven."

When positive and negative numbers are used to indicate direction of measurement, we are concerned only with absolute value, if we wish to know only the distance covered. For example, in figure 3-2, if an object moves to the left from the starting point to the point indicated by -2, the actual distance covered is 2 units. We are concerned only with the fact that $|-2| = 2$, if our only interest is in the distance and not the direction.

OPERATING WITH SIGNED NUMBERS

The number line can be used to demonstrate addition of signed numbers. Two cases must be considered; namely, adding numbers with like signs and adding numbers with unlike signs.

ADDING WITH LIKE SIGNS

As an example of addition with like signs, suppose that we use the number line (fig. 3-4) to add 2 + 3. Since these are signed numbers, we indicate this addition as (+2) + (+3). This emphasizes that, among the three + signs shown two are number signs and one is a sign of

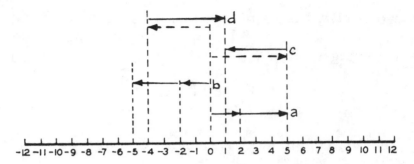

Figure 3-4.—Using the number line to add.

operation. Line a (fig. 3-4) above the number line shows this addition. Find 2 on the number line. To add 3 to it, go three units more in a positive direction and get 5.

To add two negative numbers on the number line, such as -2 and -3, find -2 on the number line and then go three units more in the negative direction to get -5, as in b (fig. 3-4) above the number line.

Observation of the results of the foregoing operations on the number line leads us to the following conclusion, which may be stated as a law: To add numbers with like signs, add the absolute values and prefix the common sign.

ADDING WITH UNLIKE SIGNS

To add a positive and a negative number, such as (-4) + (+5), find +5 on the number line and go four units in a negative direction, as in line c above the number line in figure 3-4.

Notice that this addition could be performed in the other direction. That is, we could start at -4 and move 5 units in the positive direction. (See line d, fig. 3-4.)

61

The results of our operations with mixed signs on the number line lead to the following conclusion, which may be stated as a law: To add numbers with unlike signs, find the difference between their absolute values and prefix the sign of the numerically greater number.

The following examples show the addition of the numbers 3 and 5 with the four possible combinations of signs:

3	-3	3	-3
5	-5	-5	5
8	-8	-2	2

In the first example, 3 and 5 have like signs and the common sign is understood to be positive. The sum of the absolute values is 8 and no sign is prefixed to this sum, thus signifying that the sign of the 8 is understood to be positive.

In the second example, the 3 and 5 again have like signs, but their common sign is negative. The sum of the absolute values is 8, and this time the common sign is prefixed to the sum. The answer is thus -8.

In the third example, the 3 and 5 have unlike signs. The difference between their absolute values is 2, and the sign of the larger addend is negative. Therefore, the answer is -2.

In the fourth example, the 3 and 5 again have unlike signs. The difference of the absolute values is still 2, but this time the sign of the larger addend is positive. Therefore, the sign prefixed to the 2 is positive (understood) and the final answer is simply 2.

These four examples could be written in a different form, emphasizing the distinction between the sign of a number and an operational sign, as follows:

$$(+3) + (+5) = +8$$
$$(-3) + (-5) = -8$$
$$(+3) + (-5) = -2$$
$$(-3) + (+5) = +2$$

Practice problems. Add as indicated:

1. $-10 + 5 = (-10) + (+5) = ?$
2. Add -9, -16, and 25
3. $-7 - 1 - 3 = (-7) + (-1) + (-3) = ?$
4. Add -22 and -13

Answers:

1. -5 3. -11
2. 0 4. -35

SUBTRACTION

Subtraction is the inverse of addition. When subtraction is performed, we "take away" the subtrahend. This means that whatever the value of the subtrahend, its effect is to be reversed when subtraction is indicated. In addition, the sum of 5 and -2 is 3. In subtraction, however, to take away the effect of the -2, the quantity +2 must be added. Thus the difference between +5 and -2 is +7.

Keeping this idea in mind, we may now proceed to examine the various combinations of subtraction involving signed numbers. Let us first consider the four possibilities where the minuend is numerically greater than the subtrahend, as in the following examples:

$$\begin{array}{cccc} 8 & 8 & -8 & -8 \\ \underline{5} & \underline{-5} & \underline{5} & \underline{-5} \\ 3 & 13 & -13 & -3 \end{array}$$

We may show how each of these results is obtained by use of the number line, as shown in figure 3-5.

In the first example, we find +8 on the number line, then subtract 5 by making a movement that reverses its sign. Thus, we move to the left 5 units. The result (difference) is +3. (See line a, fig. 3-5.)

In the second example, we find +8 on the number line, then subtract (-5) by making a movement that will reverse its sign. Thus we move to the right 5 units. The result in this case is +13. (See line b, fig. 3-5.)

In the third example, we find -8 on the number line, then subtract 5 by making a movement that reverses its sign. Thus we move to the left 5 units. The result is -13. (See line c, fig. 3-5.)

In the fourth example, we find -8 on the number line, then reverse the sign of -5 by moving 5 units to the right. The result is -3. (See line d, fig. 3-5.)

Next, let us consider the four possibilities that arise when the subtrahend is numerically greater than the minuend, as in the following examples:

5	5	-5	-5
8	-8	8	-8
-3	13	-13	3

In the first example, we find +5 on the number line, then subtract 8 by making a movement that reverses its sign. Thus we move to the left 8 units. The result is -3. (See line e, fig. 3-5.)

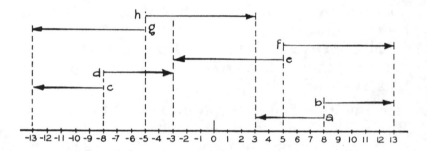

Figure 3-5.—Subtraction by use of the number line.

In the second example, we find +5 on the number line, then subtract -8 by making a movement to the right that reverses its sign. The result is 13. (See line f, fig. 3-5.)

In the third example, we find -5 on the number line, then reverse the sign of 8 by a movement to the left. The result is -13. (See line g, fig. 3-5.)

In the fourth example, we find -5 on the number line, then reverse the sign of -8 by a movement to the right. The result is 3. (See line h, fig. 3-5.)

Careful study of the preceding examples leads to the following conclusion, which is stated as a law for subtraction of signed numbers: In any subtraction problem, mentally change the sign of the subtrahend and proceed as in addition.

Practice problems. In problems 1 through 4, subtract the lower number from the upper. In 5 through 8, subtract as indicated.

1.	17	2.	-12	3.	-9	4.	7
	-10		8		-13		16

5. $1 - (-5) = ?$
6. $-6 - (-8) = ?$
7. $14 - 7 - (-3) = ?$
8. $-9 - 2 = ?$

Answers:

1. 27	2. -20	3. 4	4. -9
5. 6	6. 2	7. 10	8. -11

MULTIPLICATION

To explain the rules for multiplication of signed numbers, we recall that multiplication of whole numbers may be thought of as shortened addition. Two types of multiplication problems must be examined; the first type involves numbers with unlike signs, and the second involves numbers with like signs.

Unlike Signs

Consider the example $3(-4)$, in which the multiplicand is negative. This means we are to add -4 three times; that is, 3(-4) is equal to (-4) + (-4) + (-4), which is equal to -12. For example, if we have three 4-dollar debts, we owe 12 dollars in all.

When the multiplier is negative, as in -3(7), we are to take away 7 three times. Thus, -3(7) is equal to -(7) - (7) - (7) which is equal to -21. For example, if 7 shells were expended in one firing, 7 the next, and 7 the next, there would be a loss of 21 shells in all. Thus, the rule is as follows: The product of two numbers with unlike signs is negative.

The law of signs for unlike signs is sometimes stated as follows: Minus times plus is minus; plus times minus is minus. Thus a problem such as 3(-4) can be reduced to the following two steps:

1. Multiply the signs and write down the sign of the answer before working with the numbers themselves.

2. Multiply the numbers as if they were unsigned numbers.

Using the suggested procedure, the sign of the answer for 3(-4) is found to be minus. The product of 3 and 4 is 12, and the final answer is -12. When there are more than two numbers to be multiplied, the signs are taken in pairs until the final sign is determined.

Like Signs

When both factors are positive, as in 4(5), the sign of the product is positive. We are to add +5 four times, as follows:

$$4(5) = 5 + 5 + 5 + 5 = 20$$

When both factors are negative, as in -4(-5), the sign of the product is positive. We are to take away -5 four times.

$$-4(-5) = -(-5) - (-5) - (-5) - (-5)$$
$$= +5 \ +5 \ +5 \ +5$$
$$= 20$$

Remember that taking away a negative 5 is the same as adding a positive 5. For example, suppose someone owes a man 20 dollars and pays him back (or diminishes the debt) 5 dollars at a time. He takes away a debt of 20 dollars by giving him four positive 5-dollar bills, or a total of 20 positive dollars in all.

The rule developed by the foregoing example is as follows: The product of two numbers with like signs is positive.

Knowing that the product of two positive numbers or two negative numbers is positive, we can conclude that the product of any even number of negative numbers is positive. Similarly, the product of any odd number of negative numbers is negative.

The laws of signs may be combined as follows: Minus times plus is minus; plus times minus is minus; minus times minus is plus; plus times plus is plus. Use of this combined rule may be illustrated as follows:

$$4(-2) \cdot (-5) \cdot (6) \cdot (-3) = -720$$

Taking the signs in pairs, the understood plus

on the 4 times the minus on the 2 produces a minus. This minus times the minus on the 5 produces a plus. This plus times the understood plus on the 6 produces a plus. This plus times the minus on the 3 produces a minus, so we know that the final answer is negative. The product of the numbers, disregarding their signs, is 720; therefore, the final answer is -720.

Practice problems. Multiply as indicated:

1. $5(-8)$ = ?
2. $-7(3)(2)$ = ?
3. $6(-1)(-4)$ = ?
4. $-2(3)(-4)(5)(-6)$ = ?

Answers:

1. -40   3. 24
2. -42   4. -720

DIVISION

Because division is the inverse of multiplication, we can quickly develop the rules for division of signed numbers by comparison with the corresponding multiplication rules, as in the following examples:

1. Division involving two numbers with unlike signs is related to multiplication with unlike signs, as follows:

$$3(-4) = -12$$

Therefore, $\dfrac{-12}{3} = -4$

69

Thus, the rule for division with unlike signs is:
The quotient of two numbers with unlike signs
is negative.

2. Division involving two numbers with like
signs is related to multiplication with like signs,
as follows:

$$3(-4) = -12$$

Therefore, $\dfrac{-12}{-4} = 3$

Thus the rule for division with like signs is:
The quotient of two numbers with like signs is
positive.

The following examples show the application
of the rules for dividing signed numbers:

$$\frac{12}{3} = 4 \qquad \frac{-12}{3} = -4$$

$$\frac{-12}{-3} = 4 \qquad \frac{12}{-3} = -4$$

Practice problems. Multiply and divide as
indicated:

1. $15 \div -5$

2. $-2(-3)/-6$

3. $\dfrac{(-3)(4)}{-6}$

4. $-81/9$

Answers:

1. -3   3. 2
2. -1   4. -9

70

SPECIAL CASES

Two special cases arise frequently in which the laws of signs may be used to advantage. The first such usage is in simplifying subtraction; the second is in changing the signs of the numerator and denominator when division is indicated in the form of a fraction.

Subtraction

The rules for subtraction may be simplified by use of the laws of signs, if each expression to be subtracted is considered as being multiplied by a negative sign. For example, 4 -(-5) is the same as 4 + 5, since minus times minus is plus. This result also establishes a basis for the rule governing removal of parentheses.

The parentheses rule, as usually stated, is: Parentheses preceded by a minus sign may be removed, if the signs of all terms within the parentheses are changed. This is illustrated as follows:

$$12 -(3 - 2 + 4) = 12 - 3 + 2 - 4$$

The reason for the changes of sign is clear when the negative sign preceding the parentheses is considered to be a multiplier for the whole parenthetical expression.

Division in Fractional Form

Division is often indicated by writing the dividend as the numerator, and the divisor as the denominator, of a fraction. In algebra, every fraction is considered to have three signs. The numerator has a sign, the denominator has a sign, and the fraction itself, taken as a whole, has a sign. In many cases, one or more of these signs will be positive, and thus will not be shown. For example, in the following fraction the sign of the numerator and the sign of the denominator are both positive (understood) and the sign of the fraction itself is negative:

$$-\frac{4}{5}$$

Fractions with more than one negative sign are always reducible to a simpler form with at most one negative sign. For example, the sign of the numerator and the sign of the denominator may be both negative. We note that minus divided by minus gives the same result as plus divided by plus. Therefore, we may change to the less complicated form having plus signs (understood) for both numerator and denominator, as follows:

$$\frac{-15}{-5} = \frac{+15}{+5} = \frac{15}{5}$$

Since -15 divided by -5 is 3, and 15 divided by 5 is also 3, we conclude that the change of sign does not alter the final answer. The same reasoning may be applied in the following example, in which the sign of the fraction itself is negative:

$$-\frac{-15}{-5} = -\frac{+15}{+5} = -\frac{15}{5}$$

When the fraction itself has a negative sign, as in this example, the fraction may be enclosed in parentheses temporarily, for the purpose of working with the numerator and denominator only. Then the sign of the fraction is applied separately to the result, as follows:

$$- \frac{-15}{-5} = -\left(\frac{-15}{-5}\right) = -(3) = -3$$

All of this can be done mentally.

If a fraction has a negative sign in one of the three sign positions, this sign may be moved to another position. Such an adjustment is an advantage in some types of complicated expressions involving fractions. Examples of this type of sign change follow:

$$- \frac{15}{5} = \frac{-15}{5} = \frac{15}{-5}$$

In the first expression of the foregoing example, the sign of the numerator is positive (understood) and the sign of the fraction is negative. Changing both of these signs, we obtain the second expression. To obtain the third expression from the second, we change the sign of the numerator and the sign of the denominator. Observe that the sign changes in each case involve a pair of signs. This leads to the law of signs for fractions: Any two of the three signs of a fraction may be changed without altering the value of the fraction.

AXIOMS AND LAWS

An axiom is a self-evident truth. It is a truth that is so universally accepted that it does

73

not require proof. For example, the statement that "a straight line is the shortest distance between two points" is an axiom from plane geometry. One tends to accept the truth of an axiom without proof, because anything which is axiomatic is, by its very nature, obviously true. On the other hand, a law (in the mathematical sense) is the result of defining certain quantities and relationships and then developing logical conclusions from the definitions.

AXIOMS OF EQUALITY

The four axioms of equality with which we are concerned in arithmetic and algebra are stated as follows:

1. If the same quantity is added to each of two equal quantities, the resulting quantities are equal. This is sometimes stated as follows: If equals are added to equals, the results are equal. For example, by adding the same quantity (3) to both sides of the following equation, we obtain two sums which are equal:

$$-2 = -3 + 1$$
$$-2 + 3 = -3 + 1 + 3$$
$$1 = 1$$

2. If the same quantity is subtracted from each of two equal quantities, the resulting quantities are equal. This is sometimes stated as follows: If equals are subtracted from equals, the results are equal. For example, by subtracting 2 from both sides of the following equation we obtain results which are equal:

$$5 = 2 + 3$$
$$5 - 2 = 2 + 3 - 2$$
$$3 = 3$$

3. If two equal quantities are multiplied by the same quantity, the resulting products are equal. This is sometimes stated as follows: If equals are multiplied by equals, the products are equal. For example, both sides of the following equation are multiplied by -3 and equal results are obtained:

$$5 = 2 + 3$$
$$(-3)(5) = (-3)(2 + 3)$$
$$-15 = -15$$

4. If two equal quantities are divided by the same quantity, the resulting quotients are equal. This is sometimes stated as follows: If equals are divided by equals, the results are equal. For example, both sides of the following equation are divided by 3, and the resulting quotients are equal:

$$12 + 3 = 15$$

$$\frac{12 + 3}{3} = \frac{15}{3}$$

$$4 + 1 = 5$$

These axioms are especially useful when letters are used to represent numbers. If we know that $5x = -30$, for instance, then dividing both $5x$ and -30 by 5 leads to the conclusion that $x = -6$.

LAWS FOR COMBINING NUMBERS

Numbers are combined in accordance with the following basic laws:
1. The associative laws of addition and multiplication.
2. The commutative laws of addition and multiplication.
3. The distributive law.

Associative Law of Addition

The word "associative" suggests association or grouping. This law states that the sum of three or more addends is the same regardless of the manner in which they are grouped. For example, 6 + 3 + 1 is the same as 6 + (3 + 1) or (6 + 3) + 1 or (6 + 1) + 3.

This law can be applied to subtraction by changing signs in such a way that all negative signs are treated as number signs rather than operational signs. That is, some of the addends can be negative numbers. For example, 6 - 4 - 2 can be rewritten as 6 + (-4) + (-2). By the associative law, this is the same as

$$6 + [(-4) + (-2)] \text{ or } [6 + (-4)] + (-2).$$

However, 6 - 4 - 2 is not the same as 6 - (4 - 2); the terms must be expressed as addends before applying the associative law of addition.

Associative Law of Multiplication

This law states that the product of three or more factors is the same regardless of the manner in which they are grouped. For example, $6 \cdot 3 \cdot 2$ is the same as $(6 \cdot 3) \cdot 2$ or $6 \cdot (3 \cdot 2)$ or $(6 \cdot 2) \cdot 3$. Negative signs require no special treatment in the application of this law. For example, $6 \cdot (-4) \cdot (-2)$ is the same as $[6 \cdot (-4)] \cdot (-2)$ or $6 \cdot [(-4) \cdot (-2)]$.

Commutative Law of Addition

The word "commute" means to change, substitute or move from place to place. The commutative law of addition states that the sum of two or more addends is the same regardless of the order in which they are arranged. For example, $4 + 3 + 2$ is the same as $4 + 2 + 3$ or $2 + 4 + 3$.

This law can be applied to subtraction by changing signs so that all negative signs become number signs and all signs of operation are positive. For example, $5 - 3 - 2$ is changed to $5 + (-3) + (-2)$, which is the same as $5 + (-2) + (-3)$ or $(-3) + 5 + (-2)$.

Commutative Law of Multiplication

This law states that the product of two or more factors is the same regardless of the order in which the factors are arranged. For example, $3 \cdot 4 \cdot 5$ is the same as $5 \cdot 3 \cdot 4$ or $4 \cdot 3 \cdot 5$. Negative signs require no special treatment in the application of this law. For example, $2 \cdot (-4) \cdot (-3)$ is the same as $(-4) \cdot (-3) \cdot 2$ or $(-3) \cdot 2 \cdot (-4)$.

Distributive Law

This law combines the operations of addition

and multiplication. The word "distributive" refers to the distribution of a common multiplier among the terms of an additive expression. For example,

$$2(3 + 4 + 5) = 2 \cdot 3 + 2 \cdot 4 + 2 \cdot 5$$
$$= 6 + 8 + 10$$

To verify the distributive law, we note that $2(3 + 4 + 5)$ is the same as $2(12)$ or 24. Also, $6 + 8 + 10$ is 24. For application of the distributive law where negative signs appear, the following procedure is recommended:

$$3(4 - 2) = 3 \left[4 + (-2) \right]$$
$$= 3(4) + 3(-2)$$
$$= 12 - 6$$
$$= 6$$

CHAPTER 4

COMMON FRACTIONS

The emphasis in previous chapters of this course has been on integers (whole numbers). In this chapter, we turn our attention to numbers which are not integers. The simplest type of number other than an integer is a COMMON FRACTION. Common fractions and integers together comprise a set of numbers called the RATIONAL NUMBERS; this set is a subset of the set of real numbers.

The number line may be used to show the relationship between integers and fractions. For example, if the interval between 0 and 1 is marked off to form three equal spaces (thirds), then each space so formed is one-third of the total interval. If we move along the number line from 0 toward 1, we will have covered two of the three "thirds" when we reach the second mark. Thus the position of the second mark represents the number 2/3. (See fig. 4-1.)

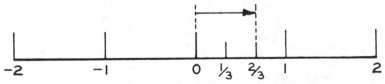

Figure 4-1.—Integers and fractions on the number line.

79

The numerals 2 and 3 in the fraction 2/3 are named so that we may distinguish between them; 2 is the NUMERATOR and 3 is the DENOMINA-TOR. In general, the numeral above the dividing line in a fraction is the numerator and the numeral below the line is the denominator. The numerator and denominator are the TERMS of the fraction. The word "numerator" is related to the word "enumerate." To enumerate means to "tell how many"; thus the numerator tells us how many fractional parts we have in the indicated fraction. To denominate means to "give a name" or "tell what kind"; thus the denominator tells us what kind of parts we have (halves, thirds, fourths, etc.).

Attempts to define the word "fraction" in mathematics usually result in a statement similar to the following: A fraction is an indicated division. Any division may be indicated by placing the dividend over the divisor and drawing a line between them. By this definition, any number which can be written as the ratio of two integers (one integer over the other) can be considered as a fraction. This leads to a further definition: Any number which can be expressed as the ratio of two integers is a RATIONAL number. Notice that every integer is a rational number, because we can write any integer as the numerator of a fraction having 1 as its denominator. For example, 5 is the same as 5/1. It should be obvious from the definition that every common fraction is also a rational number.

TYPES OF FRACTIONS

Fractions are often classified as proper or

improper. A proper fraction is one in which the numerator is numerically smaller than the denominator. An improper fraction has a numerator which is larger than its denominator.

MIXED NUMBERS

When the denominator of an improper fraction is divided into its numerator, a remainder is produced along with the quotient, unless the numerator happens to be an exact multiple of the denominator. For example, 7/5 is equal to 1 plus a remainder of 2. This remainder may be shown as a dividend with 5 as its divisor, as follows:

$$\frac{7}{5} = \frac{5 + 2}{5} = 1 + \frac{2}{5}$$

The expression 1 + 2/5 is a MIXED NUMBER. Mixed numbers are usually written without showing the plus sign; that is, 1 + 2/5 is the same as $1\frac{2}{5}$ or 1 2/5. When a mixed number is written as 1 2/5, care must be taken to insure that there is a space between the 1 and the 2; otherwise, 1 2/5 might be taken to mean 12/5.

MEASUREMENT FRACTIONS

Measurement fractions occur in problems such as the following:

If $2 were spent for a stateroom rug at $3 per yard, how many yards were bought? If $6 had been spent we could find the number of yards by simply dividing the cost per yard into the amount spent. Since 6/3 is 2, two yards could be bought for $6. The same reasoning applies when $2 are spent, but in this case we can only indicate the amount purchased as the indicated division 2/3. Figure 4-2 shows a diagram for both the $6 purchase and the $2 purchase.

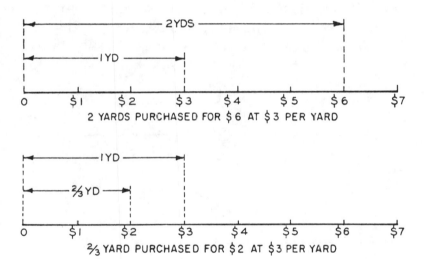

Figure 4-2.—Measurement fractions.

PARTITIVE FRACTIONS

The difference between measurement fractions and partitive fractions is explained as follows: Measurement fractions result when we determine how many pieces of a given size can be cut from a larger piece. Partitive fractions result when we cut a number of pieces of equal size from a larger piece and then determine the size of each smaller piece. For example, if 4 equal lengths of pipe are to be cut from a 3-foot pipe, what is the size of each piece? If the problem had read that 3 equal lengths were to be cut from a 6-foot pipe, we could find the size of each pipe by dividing the number of equal lengths into the overall length. Thus, since 6/3 is 2, each piece would be 2 feet long. By this same reasoning in the example, we divide the overall length by the number of equal parts to get the size of the individual pieces; that is, 3/4 foot. The partitioned 6-foot and 3-foot pipes are shown in figure 4-3.

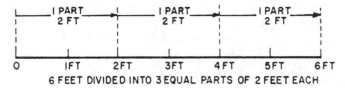

6 FEET DIVIDED INTO 3 EQUAL PARTS OF 2 FEET EACH

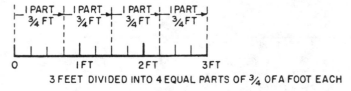

3 FEET DIVIDED INTO 4 EQUAL PARTS OF 3/4 OF A FOOT EACH

Figure 4-3.—Partitive fractions

EXPRESSING RELATIONSHIPS

When a fraction is used to express a relationship, the numerator and denominator take on individual significance. In this frame of reference, 3/4 means 3 out of 4, or 3 parts in 4, or the ratio of 3 to 4. For example, if 1 out of 3 of the men in a division are on liberty, then it would be correct to state that 1/3 of the division are on liberty. Observe that neither of these ways of expressing the relationship tells us the actual number of men; the relationship itself is the important thing.

Practice problems.

1. What fraction of 1 foot is 11 inches?
2. Represent 3 out of 8 as a fraction.
3. Write the fractions that indicate the relationship of 2 to 3; 8 divided by 9; and 6 out of 7 equal parts.
4. The number $6\frac{3}{5}$ means $6 \underline{} \frac{3}{5}$

Answers:

1. 11/12
2. 3/8
3. 2/3; 8/9; 6/7
4. plus

EQUIVALENT FRACTIONS

It will be recalled that any number divided by itself is 1. For example, 1/1, 2/2, 3/3, 4/4, and all other numbers formed in this way, have the value 1. Furthermore, any number multiplied by 1 is equivalent to the number itself. For example, 1 times 2 is 2, 1 times 3 is 3, 1 times 1/2 is 1/2, etc.

These facts are used in changing the form of a fraction to an equivalent form which is more convenient for use in a particular problem. For example, if 1 in the form $\frac{2}{2}$ is multiplied by

$\frac{3}{5}$, the product will still have a value of $\frac{3}{5}$ but will be in a different form, as follows:

$$\frac{2}{2} \cdot \frac{3}{5} = \frac{2 \cdot 3}{2 \cdot 5} = \frac{6}{10}$$

Figure 4-4 shows that $\frac{3}{5}$ of line a is equal to $\frac{6}{10}$ of line b where line a equals line b. Line a is marked off in fifths and line b is marked off in tenths. It can readily be seen that $\frac{6}{10}$ and $\frac{3}{5}$ measure distances of equal length.

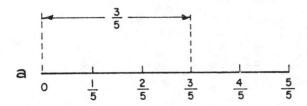

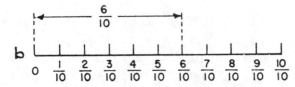

Figure 4-4.—Equivalent fractions.

The markings on a ruler show equivalent fractions. The major division of an inch divides it into two equal parts. One of these parts represents $\frac{1}{2}$. The next smaller markings divide the inch into four equal parts. It will be noted that two of these parts represent the same distance as

85

$\frac{1}{2}$; that is, $\frac{2}{4}$ equals $\frac{1}{2}$. Also, the next smaller markings break the inch into 8 equal parts. How many of these parts are equivalent to $\frac{1}{2}$ inch?

The answer is found by noting that $\frac{4}{8}$ equals $\frac{1}{2}$.

Practice problems. Using the divisions on a ruler for reference, complete the following exercise:

1. $\frac{1}{4} = \frac{?}{8}$

2. $\frac{1}{8} = \frac{?}{16}$

3. $\frac{3}{4} = \frac{?}{16}$

4. $\frac{1}{4} = \frac{?}{16}$

Answers:

1. 2

2. 2

3. 12

4. 4

A review of the foregoing exercise will reveal that in each case the right-hand fraction could be formed by multiplying both the numerator and the denominator of the left-hand fraction by the same number. In each case the number may be determined by dividing the denominator of the right-hand fraction by the denominator of the left-hand fraction. Thus in problem 1, both terms of $\frac{1}{4}$ were multiplied by 2.

In problem 3, both terms were multiplied by 4. It is seen that multiplying both terms of a fraction by the same number does not change the value of the fraction.

Since $\frac{1}{2}$ equals $\frac{2}{4}$, the reverse must also be true; that is $\frac{2}{4}$ must be equal to $\frac{1}{2}$. This can likewise be verified on a ruler. We have already seen that $\frac{4}{8}$ is the same as $\frac{1}{2}$, $\frac{12}{16}$ equals $\frac{3}{4}$, and $\frac{2}{8}$ equals $\frac{1}{4}$. We see that dividing both terms of a fraction by the same number does not change the value of the fraction.

FUNDAMENTAL RULE OF FRACTIONS

The foregoing results are combined to form the fundamental rule of fractions, which is stated as follows: Multiplying or dividing both terms of a fraction by the same number does not change the value of the fraction. This is one of the most important rules used in dealing with fractions.

The following examples show how the fundamental rule is used:

1. Change 1/4 to twelfths. This problem is set up as follows:

$$\frac{1}{4} = \frac{?}{12}$$

The first step is to determine how many 4's are contained in 12. The answer is 3, so we know that the multiplier for both terms of the fraction is 3, as follows:

$$\frac{3}{3} \cdot \frac{1}{4} = \frac{3}{12}$$

2. What fraction with a numerator of 6 is equal to 3/4?

SOLUTION:
$$\frac{6}{?} = \frac{3}{4}$$

We note that 6 contains 3 twice; therefore we need to double the numerator of the right-hand fraction to make it equivalent to the numerator of the fraction we seek. We multiply both terms of 3/4 by 2, obtaining 8 as the denominator of the new fraction, as follows:

$$\frac{6}{8} = \frac{3}{4} \cdot \frac{2}{2}$$

3. Change 6/16 to eighths.

SOLUTION:
$$\frac{6}{16} = \frac{?}{8}$$

We note that the denominator of the fraction which we seek is 1/2 as large as the denominator of the original fraction. Therefore the new fraction may be formed by dividing both terms of the original fraction by 2, as follows:

$$\frac{6 \div 2}{16 \div 2} = \frac{3}{8}$$

Practice problems. Supply the missing number in each of the following:

1. $\frac{3}{8} = \frac{30}{?}$ 3. $\frac{?}{90} = \frac{3}{10}$ 5. $\frac{1}{?} = \frac{12}{72}$

2. $\frac{44}{48} = \frac{?}{12}$ 4. $\frac{1}{6} = \frac{6}{?}$ 6. $\frac{3}{5} = \frac{?}{25}$

88

Answers:

1. 80   3. 27   5. 6
2. 11   4. 36   6. 15

REDUCTION TO LOWEST TERMS

It is frequently desirable to change a fraction to an equivalent fraction with the smallest possible terms; that is, with the smallest possible numerator and denominator. This process is called REDUCTION. Thus, $\frac{6}{30}$ reduced to lowest terms is $\frac{1}{5}$. Reduction can be accomplished by finding the largest factor that is common to both the numerator and denominator and dividing both of these terms by it. Dividing both terms of the preceding example by 6 reduces the fraction to lowest terms. In computation, fractions should usually be reduced to lowest terms where possible.

If the greatest common factor cannot readily be found, any common factor may be removed and the process repeated until the fraction is in lowest terms: Thus, $\frac{18}{48}$ could first be divided by 2 and then by 3.

$$\frac{18 \div 2}{48 \div 2} = \frac{9}{24}$$

$$\frac{9 \div 3}{24 \div 3} = \frac{3}{8}$$

Practice problems. Reduce the following fractions to lowest terms:

89

1. $\dfrac{18}{48}$ 2. $\dfrac{15}{20}$ 3. $\dfrac{35}{56}$

4. $\dfrac{12}{60}$ 5. $\dfrac{18}{24}$ 6. $\dfrac{9}{144}$

Answers:

1. $\dfrac{3}{8}$ 2. $\dfrac{3}{4}$ 3. $\dfrac{5}{8}$

4. $\dfrac{1}{5}$ 5. $\dfrac{3}{4}$ 6. $\dfrac{1}{16}$

IMPROPER FRACTIONS

Although the "improper" fraction is really quite "proper" mathematically, it is usually customary to change it to a mixed number. A recipe may call for $1\frac{1}{2}$ cups of milk, but would not call for $\frac{3}{2}$ cups of milk.

Since a fraction is an indicated division, a method is already known for reduction of improper fractions to mixed numbers. The improper fraction $\frac{8}{3}$ may be considered as the division of 8 by 3. This division is carried out as follows:

$$\begin{array}{r} 2\ R\ 2 = 2\frac{2}{3} \\ 3\overline{)8} \\ \underline{6} \\ 2 \end{array}$$

The truth of this can be verified another way

90

If 1 equals $\frac{3}{3}$, then 2 equals $\frac{6}{3}$. Thus,

$$2\frac{2}{3} = 2 + \frac{2}{3} = \frac{6}{3} + \frac{2}{3} = \frac{8}{3}$$

These examples lead to the following conclusion, which is stated as a rule: To change an improper fraction to a mixed number, divide the numerator by the denominator and write the fractional part of the quotient in lowest terms.

Practice problems. Change the following fractions to mixed numbers:

1. 31/20   3. 65/20
2. 33/9   4. 45/8

Answers:

1. $1\frac{11}{20}$ 3. $3\frac{1}{4}$

2. $3\frac{2}{3}$ 4. $5\frac{5}{8}$

OPERATING WITH MIXED NUMBERS

In computation, mixed numbers are often unwieldy. As it is possible to change any improper fraction to a mixed number, it is likewise possible to change any mixed number to an improper fraction. The problem can be reduced to the finding of an equivalent fraction and a simple addition.

EXAMPLE: Change $2\frac{1}{5}$ to an improper fraction.

91

SOLUTION:

Step 1: Write $2\frac{1}{5}$ as a whole number plus a fraction, $2 + \frac{1}{5}$.

Step 2: Change 2 to an equivalent fraction with a denominator of 5, as follows:

$$\frac{2}{1} = \frac{?}{5}$$

$$\frac{2(5)}{1(5)} = \frac{10}{5}$$

Step 3: Add $\frac{10}{5} + \frac{1}{5} = \frac{11}{5}$

Thus, $\qquad 2\frac{1}{5} = \frac{11}{5}$

EXAMPLE: Write $5\frac{2}{9}$ as an improper fraction.

SOLUTION: $\qquad 5\frac{2}{9} = 5 + \frac{2}{9}$

$$\frac{5}{1} = \frac{?}{9}$$

$$\frac{5(9)}{1(9)} = \frac{45}{9}$$

$$\frac{45}{9} + \frac{2}{9} = \frac{47}{9}$$

Thus, $\qquad 5\frac{2}{9} = \frac{47}{9}$

In each of these examples, notice that the multiplier used in step 2 is the same number as the denominator of the fractional part of the original mixed number. This leads to the following conclusion, which is stated as a rule:

To change a mixed number to an improper fraction, multiply the whole-number part by the denominator of the fractional part and add the numerator to this product. The result is the numerator of the improper fraction; its denominator is the same as the denominator of the fractional part of the original mixed number.

Practice problems. Change the following mixed numbers to improper fractions:

1. $1\frac{1}{5}$ 　　　　　 3. $3\frac{2}{7}$

2. $2\frac{11}{20}$ 　　　　 4. $4\frac{3}{10}$

Answers:

1. $\frac{6}{5}$ 　　　　　 3. $\frac{23}{7}$

2. $\frac{51}{20}$ 　　　　 4. $\frac{43}{10}$

NEGATIVE FRACTIONS

A fraction preceded by a minus sign is negative. Any negative fraction is equivalent to a positive fraction multiplied by -1. For example,

$$-\frac{2}{5} = -1\left(\frac{2}{5}\right)$$

93

The number $-\dfrac{2}{5}$ is read "minus two-fifths."

We know that the quotient of two numbers with unlike signs is negative. Therefore,

$$\dfrac{-2}{5} = -\dfrac{2}{5} \text{ and } \dfrac{2}{-5} = -\dfrac{2}{5}$$

This indicates that a negative fraction is equivalent to a fraction with either a negative numerator or a negative denominator.

The fraction $\dfrac{2}{-5}$ is read "two over minus five." The fraction $\dfrac{-2}{5}$ is read "minus two over five."

A minus sign in a fraction can be moved about at will. It can be placed before the numerator, before the denominator, or before the fraction itself. Thus,

$$\dfrac{-2}{5} = \dfrac{2}{-5} = -\dfrac{2}{5}$$

Moving the minus sign from numerator to denominator, or vice versa, is equivalent to multiplying the terms of the fraction by -1. This is shown in the following examples:

$$\dfrac{-2(-1)}{5(-1)} = \dfrac{2}{-5} \text{ and } \dfrac{2(-1)}{-5(-1)} = \dfrac{-2}{5}$$

A fraction may be regarded as having three signs associated with it—the sign of the numerator, the sign of the denominator, and the sign preceding the fraction. Any two of these signs may be changed without changing the value of the fraction. Thus,

$$-\frac{3}{4} = \frac{-3}{4} = \frac{3}{-4} = -\frac{-3}{-4}$$

OPERATIONS WITH FRACTIONS

It will be recalled from the discussion of denominate numbers that numbers must be of the same denomination to be added. We can add pounds to pounds, pints to pints, but not ounces to pints. If we think of fractions loosely as denominate numbers, it will be seen that the rule of likeness applies also to fractions. We can add eighths to eighths, fourths to fourths, but not eighths to fourths. To add $\frac{1}{5}$ inch to $\frac{2}{5}$ inch we simply add the numerators and retain the denominator unchanged. The denomination is fifths; as with denominate numbers, we add 1 fifth to 2 fifths to get 3 fifths, or $\frac{3}{5}$.

LIKE AND UNLIKE FRACTIONS

We have shown that like fractions are added by simply adding the numerators and keeping the denominator. Thus,

$$\frac{3}{8} + \frac{2}{8} = \frac{3 + 2}{8} = \frac{5}{8}$$

or

$$\frac{5}{16} + \frac{2}{16} = \frac{7}{16}$$

Similarly we can subtract like fractions by subtracting the numerators.

$$\frac{7}{8} - \frac{2}{8} = \frac{7 - 2}{8} = \frac{5}{8}$$

The following examples will show that like fractions may be divided by dividing the numerator of the dividend by the numerator of the divisor.

$$\frac{3}{8} \div \frac{1}{8} = ?$$

SOLUTION: We may state the problem as a question: "How many times does $\frac{1}{8}$ appear in $\frac{3}{8}$, or how many times may $\frac{1}{8}$ be taken from $\frac{3}{8}$?"

$$3/8 - 1/8 = 2/8 \qquad (1)$$
$$2/8 - 1/8 = 1/8 \qquad (2)$$
$$1/8 - 1/8 = 0/8 = 0 \qquad (3)$$

We see that 1/8 can be subtracted from 3/8 three times. Therefore,

$$3/8 \div 1/8 = 3$$

When the denominators of fractions are unequal, the fractions are said to be unlike. Addition, subtraction, or division cannot be performed directly on unlike fractions. The proper application of the fundamental rule, however, can change their form so that they become like fractions; then all the rules for like fractions apply.

LOWEST COMMON DENOMINATOR

To change unlike fractions to like fractions, it is necessary to find a COMMON DENOMINA-TOR and it is usually advantageous to find the LOWEST COMMON DENOMINATOR (L C D). This is nothing more than the least common multiple of the denominators.

Least Common Multiple

If a number is a multiple of two or more different numbers, it is called a COMMON MULTIPLE. Thus, 24 is a common multiple of 6 and 2. There are many common multiples of these numbers. The numbers 36, 48, and 54, to name a few, are also common multiples of 6 and 2.

The smallest of the common multiples of a set of numbers is called the LEAST COMMON MULTIPLE. It is abbreviated LCM. The least common multiple of 6 and 2 is 6. To find the least common multiple of a set of numbers, first separate each of the numbers into prime factors.

Suppose that we wish to find the LCM of 14, 24, and 30. Separating these numbers into prime factors we have

$$14 = 2 \cdot 7$$
$$24 = 2^3 \cdot 3$$
$$30 = 2 \cdot 3 \cdot 5$$

The LCM will contain each of the various prime factors shown. Each prime factor is used the greatest number of times that it occurs in any one of the numbers. Notice that 3, 5, and 7 each occur only once in any one number. On the other hand, 2 occurs three times in one number. We get the following result:

$$LCM = 2^3 \cdot 3 \cdot 5 \cdot 7$$
$$= 840$$

Thus, 840 is the least common multiple of 14, 24, and 30.

Greatest Common Divisor

The largest number that can be divided into each of two or more given numbers without a remainder is called the GREATEST COMMON DIVISOR of the given numbers. It is abbreviated GCD. It is also sometimes called the HIGHEST COMMON FACTOR.

In finding the GCD of a set of numbers, separate the numbers into prime factors just as for LCM. The GCD is the product of only those factors that appear in all of the numbers. Notice in the example of the previous section that 2 is the greatest common divisor of 14, 24, and 30.

Find the GCD of 650, 900, and 700. The procedure is as follows:

$$650 = 2 \cdot 5^2 \cdot 13$$
$$900 = 2^2 \cdot 3^2 \cdot 5^2$$
$$700 = 2^2 \cdot 5^2 \cdot 7$$
$$GCD = 2 \cdot 5^2 = 50$$

Notice that 2 and 5^2 are factors of each number. The greatest common divisor is 2 x 25 = 50.

USING THE LCD

Consider the example

$$\frac{1}{2} + \frac{1}{3}$$

The numbers 2 and 3 are both prime; so the LCD is 6.

Therefore $\qquad \frac{1}{2} = \frac{3}{6}$

and $\qquad \frac{1}{3} = \frac{2}{6}$

Thus, the addition of $\frac{1}{2}$ and $\frac{1}{3}$ is performed as follows:

$$\frac{1}{2} + \frac{1}{3} = \frac{3}{6} + \frac{2}{6} = \frac{5}{6}$$

In the example

$$\frac{1}{5} + \frac{3}{10}$$

10 is the LCD.

Therefore, $\qquad \frac{1}{5} + \frac{3}{10} = \frac{2}{10} + \frac{3}{10}$

$$= \frac{5}{10} = \frac{1}{2}$$

99

Practice problems. Change the fractions in each of the following groups to like fractions with least common denominators:

1. $\dfrac{1}{3}, \dfrac{1}{6}$

2. $\dfrac{5}{12}, \dfrac{2}{3}$

3. $\dfrac{1}{2}, \dfrac{1}{4}, \dfrac{2}{3}$

4. $\dfrac{1}{6}, \dfrac{3}{10}, \dfrac{1}{5}$

Answers:

1. $\dfrac{2}{6}, \dfrac{1}{6}$

2. $\dfrac{5}{12}, \dfrac{8}{12}$

3. $\dfrac{6}{12}, \dfrac{3}{12}, \dfrac{8}{12}$

4. $\dfrac{5}{30}, \dfrac{9}{30}, \dfrac{6}{30}$

ADDITION

It has been shown that in adding like fractions we add the numerators. In adding unlike fractions, the fractions must first be changed so that they have common denominators. We apply these same rules in adding mixed numbers. It will be remembered that a mixed number is an indicated sum. Thus, $2\dfrac{1}{3}$ is really $2 + \dfrac{1}{3}$. Adding can be done in any order. The following examples will show the application of these rules:

EXAMPLE:

$$2\dfrac{1}{3}$$
$$3\dfrac{1}{3}$$
$$\overline{5\dfrac{2}{3}}$$

This could have been written as follows:

$$2 + \frac{1}{3}$$

$$3 + \frac{1}{3}$$

$$\overline{5 + \frac{2}{3}} = 5\frac{2}{3}$$

EXAMPLE:

$$4\frac{5}{7}$$

$$6\frac{3}{7}$$

$$\overline{10\frac{8}{7}}$$

Here we change $\frac{8}{7}$ to the mixed number $1\frac{1}{7}$. Then

$$10\frac{8}{7} = 10 + 1 + \frac{1}{7}$$

$$= 11\frac{1}{7}$$

EXAMPLE:

Add
$$\frac{1}{4}$$

$$\overline{2\frac{2}{3}}$$

101

We first change the fractions so that they are like and have the least common denominator and then proceed as before.

$$\frac{1}{4} = \frac{3}{12}$$

$$2\,\frac{2}{3} = 2\,\frac{8}{12}$$

$$\overline{\phantom{2\,\frac{2}{3} = }\ 2\,\frac{11}{12}}$$

EXAMPLE:

Add

$$4\,\frac{5}{8} = 4\,\frac{5}{8}$$

$$2\,\frac{1}{2} = 2\,\frac{4}{8}$$

$$\frac{1}{4} = \frac{2}{8}$$

$$\overline{\phantom{2\,\frac{1}{2} = }\ 6\,\frac{11}{8}}$$

Since $\frac{11}{8}$ equals $1\,\frac{3}{8}$, the final answer is found as follows:

$$6\,\frac{11}{8} = 6 + 1 + \frac{3}{8}$$

$$= 7\,\frac{3}{8}$$

Practice problems. Add, and reduce the sums to simplest terms:

1. $1\frac{1}{7}$ 2. $\frac{3}{4}$ 3. $6\frac{2}{5}$ 4. $\frac{5}{8}$ 5. $4\frac{1}{2}$

$\underline{2\frac{3}{4}}$ $\underline{1\frac{1}{2}}$ $\underline{3\frac{1}{4}}$ $\underline{2\frac{3}{20}}$ $\underline{1\frac{1}{8}}$

Answers:

1. $3\frac{25}{28}$ 2. $2\frac{1}{4}$ 3. $9\frac{13}{20}$ 4. $2\frac{31}{40}$ 5. $5\frac{5}{8}$

The following example demonstrates a practical application of addition of fractions:

EXAMPLE: Find the total length of the piece of metal shown in figure 4-5 (A).

SOLUTION: First indicate the sum as follows:

$$\frac{9}{16} + \frac{3}{4} + \frac{7}{8} + \frac{3}{4} + \frac{9}{16} = ?$$

Changing to like fractions and adding numerators,

$$\frac{9}{16} + \frac{12}{16} + \frac{14}{16} + \frac{12}{16} + \frac{9}{16} = \frac{56}{16}$$

$$= 3\frac{8}{16}$$

$$= 3\frac{1}{2}$$

The total length is $3\frac{1}{2}$ inches.

Practice problem. Find the distance from

the center of the first hole to the center of the
last hole in the metal plate shown in figure
4-5 (B).

Answer: $2\ \dfrac{7}{16}$ inches

SUBTRACTION

The rule of likeness applies in the subtraction of fractions as well as in addition.
Some examples will show that cases likely to
arise may be solved by use of ideas previously
developed.

(A)

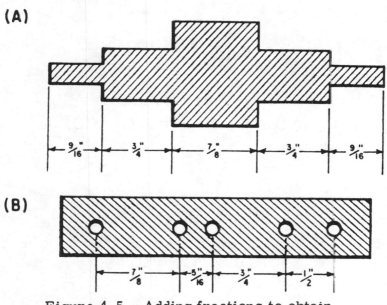

(B)

Figure 4-5.—Adding fractions to obtain
total length or spacing.

EXAMPLE: Subtract $1\dfrac{1}{3}$ from $5\ \dfrac{2}{3}$

$$5\,\frac{2}{3}$$

$$1\,\frac{1}{3}$$

$$\overline{\rule{2cm}{0pt}}$$

$$4\,\frac{1}{3}$$

We see that whole numbers are subtracted from whole numbers; fractions from fractions.

EXAMPLE: Subtract $\frac{1}{8}$ from $\frac{4}{5}$

$$\frac{4}{5}$$

$$\frac{1}{8}$$

$$\overline{\rule{2cm}{0pt}}$$

Changing to like fractions with an LCD, we have

$$\frac{32}{40}$$

$$\frac{5}{40}$$

$$\overline{\rule{2cm}{0pt}}$$

$$\frac{27}{40}$$

EXAMPLE: Subtract $\frac{11}{12}$ from $3\,\frac{2}{3}$

$$3\,\frac{2}{3} = 3\,\frac{8}{12}$$

105

$$\frac{11}{12} = \frac{11}{12}$$

Regrouping $3\frac{8}{12}$ we have

$$2 + 1 + \frac{8}{12} = 2 + \frac{12}{12} + \frac{8}{12}$$

Then

$$3\frac{2}{3} = 2\frac{20}{12}$$

$$\frac{11}{12} = \frac{11}{12}$$

$$2\frac{9}{12} = 2\frac{3}{4}$$

Practice problems. Subtract the lower number from the upper number and reduce the difference to simplest terms:

1. $\dfrac{7}{9}$ 2. $\dfrac{2}{3}$ 3. $5\dfrac{5}{12}$ 4. 5 5. $2\dfrac{3}{8}$

$\dfrac{1}{6}$ $\dfrac{1}{3}$ $2\dfrac{7}{12}$ $2\dfrac{2}{3}$ $\dfrac{5}{8}$

Answers:

1. $\dfrac{11}{18}$ 2. $\dfrac{1}{3}$ 3. $2\dfrac{5}{6}$ 4. $2\dfrac{1}{3}$ 5. $1\dfrac{3}{4}$

The following problem demonstrates subtraction of fractions in a practical situation.

EXAMPLE: What is the length of the dimension marked X on the machine bolt shown in figure 4-6 (A)?

SOLUTION: Total the lengths of the known parts.

$$\frac{1}{4} + \frac{1}{64} + \frac{1}{2} = \frac{16}{64} + \frac{1}{64} + \frac{32}{64} = \frac{49}{64}$$

Subtract this sum from the overall length.

$$2 = 1\,\frac{64}{64}$$

$$\frac{49}{64} = \frac{49}{64}$$

$$\overline{\qquad\qquad 1\,\frac{15}{64}}$$

The answer is $1\,\frac{15}{64}$ inch.

(A)

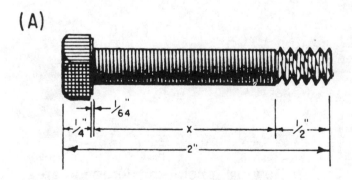

(B)

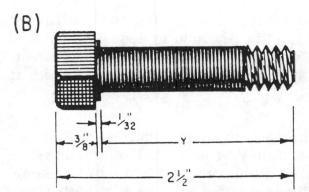

Figure 4-6.—Finding unknown dimensions
by subtracting fractions.

Practice problem. Find the length of the dimension marked Y on the machine bolt in figure 4-6 (B).

MULTIPLICATION

The fact that multiplication by a fraction does not increase the value of the product may confuse those who remember the definition of multiplication presented earlier for whole numbers. It was stated that 4(5) means 5 is taken as an addend 4 times. How is it then that $\frac{1}{2}$(4) is 2, a number less than 4? Obviously our idea of multiplication must be broadened.

Consider the following products:

$$4(4) = 16$$
$$3(4) = 12 \qquad \frac{1}{2}(4) = 2$$
$$2(4) = 8$$
$$1(4) = 4 \qquad \frac{1}{4}(4) = 1$$

Notice that as the multiplier decreases, the

product decreases, until, when the multiplier is a fraction, the product is less than 4 and continues to decrease as the fraction decreases. The fraction introduces the "part of" idea: $\frac{1}{2}(4)$ means $\frac{1}{2}$ of 4; $\frac{1}{4}(4)$ means $\frac{1}{4}$ of 4.

The definition of multiplication stated for whole numbers may be extended to include fractions. Since 4(5) means that 5 is to be used 4 times as an addend, we can say that with fractions the numerator of the multiplier tells how many times the numerator of the multiplicand is to be used as an addend. By the same reasoning, the denominator of the multiplier tells how many times the denominator of the multiplicand is to be used as an addend. The following examples illustrate the use of this idea:

1. The fraction $\frac{1}{12}$ is multiplied by the whole number 4 as follows:

$$4 \times \frac{1}{12} = \frac{4}{1} \times \frac{1}{12}$$

$$= \frac{1 + 1 + 1 + 1}{12}$$

$$= \frac{4}{12} = \frac{1}{3}$$

This example shows that 4 (1/12) is the same as $\frac{4(1)}{12}$.

Another way of thinking about the multiplication of 1/12 by 4 is as follows:

$$4 \times \frac{1}{12} = \frac{1}{12} + \frac{1}{12} + \frac{1}{12} + \frac{1}{12}$$

$$= \frac{4}{12} = \frac{1}{3}$$

2. The fraction 2/3 is multiplied by 1/2 as follows:

$$\frac{1}{2} \times \frac{2}{3} = \frac{2}{6}$$

$$= \frac{1}{3}$$

From these examples a general rule is developed: To find the product of two or more fractions multiply their numerators together and write the result as the numerator of the product; multiply their denominators and write the result as the denominator of the product; reduce the answer to lowest terms.

In using this rule with whole numbers, write each whole number as a fraction with 1 as the denominator. For example, multiply 4 times 1/12 as follows:

$$4 \times \frac{1}{12} = \frac{4}{1} \times \frac{1}{12}$$

$$= \frac{4}{12} = \frac{1}{3}$$

In using this rule with mixed numbers, rewrite all mixed numbers as improper fractions before applying the rule, as follows:

110

$$2\frac{1}{3} \times \frac{1}{2} = \frac{7}{3} \times \frac{1}{2}$$

$$= \frac{7}{6}$$

A second method of multiplying mixed numbers makes use of the distributive law. This law states that a multiplier applied to a two-part expression is distributed over both parts. For example, to multiply $6\frac{1}{3}$ by 4 we may rewrite $6\frac{1}{3}$ as 6 + 1/3. Then the problem can be written as 4(6 + 1/3) and the multiplication proceeds as follows:

$$4(6 + 1/3) = 24 + 4/3$$
$$= 25 + 1/3$$
$$= 25\frac{1}{3}$$

Cancellation

Computation can be considerably reduced by dividing out (CANCELLING) factors common to both the numerator and the denominator. We recognize a fraction as an indicated division. Thinking of $\frac{6}{9}$ as an indicated division, we remember that we can simplify division by showing both dividend and divisor as the indicated products of their factors and then dividing like factors, or canceling. Thus,

$$\frac{6}{9} = \frac{2 \times 3}{3 \times 3}$$

Dividing the factor 3 in the numerator by 3 in the denominator gives the following simplified result:

$$\frac{2 \times \overset{1}{\cancel{3}}}{3 \times \underset{1}{\cancel{3}}} = \frac{2}{3}$$

This method is most advantageous when done before any other computation. Consider the example,

$$\frac{1}{3} \times \frac{3}{2} \times \frac{2}{5}$$

The product in factored form is

$$\frac{1 \times 3 \times 2}{3 \times 2 \times 5}$$

Rather than doing the multiplying and then reducing the result $\frac{6}{30}$, it is simpler to cancel like factors first, as follows:

$$\frac{1 \times \overset{1}{\cancel{3}} \times \overset{1}{\cancel{2}}}{\underset{1}{\cancel{3}} \times \underset{1}{\cancel{2}} \times 5} = \frac{1}{5}$$

Likewise,

$$\frac{\overset{1}{\cancel{\underset{1}{\cancel{2}}}}}{\underset{1}{\cancel{3}}} \times \frac{\overset{\overset{1}{\cancel{2}}}{\cancel{6}}}{\underset{\underset{1}{\cancel{2}}}{\cancel{4}}} \times \frac{5}{9} = \frac{5}{9}$$

Here we mentally factor 6 to the form 3 x 2, and 4 to the form 2 x 2. Cancellation is a valuable tool in shortening operations with fractions.

The general rule may be applied to mixed numbers by simply changing them to improper fractions.

Thus,

$$2\frac{1}{4} \times 3\frac{1}{3} = \ ?$$

$$\frac{9}{4} \times \frac{10}{3} = \frac{\overset{3}{\cancel{9}}}{\underset{2}{\cancel{4}}} \times \frac{\overset{5}{\cancel{10}}}{\underset{1}{\cancel{3}}} = \frac{15}{2}$$

$$= 7\frac{1}{2}$$

Practice problems. Determine the following products, using the general rule and canceling where possible:

1. $\dfrac{5}{8} \times 12$

2. $\dfrac{1}{2} \times \dfrac{1}{3} \times \dfrac{2}{5}$

3. $5 \times \dfrac{4}{9}$

4. $\dfrac{3}{4} \times 6$

5. $\dfrac{1}{3} \times \dfrac{2}{3}$

6. $\dfrac{4}{3} \times \dfrac{1}{6}$

Answers:

1. $7\frac{1}{2}$ 3. $2\frac{2}{9}$ 5. $\frac{2}{9}$

2. $\frac{1}{15}$ 4. $4\frac{1}{2}$ 6. $\frac{2}{9}$

The following problem illustrates the multiplication of fractions in a practical situation.

EXAMPLE: Find the distance between the center lines of the first and fifth rivets connecting the two metal plates shown in figure 4-7 (A).

SOLUTION: The distance between two adjacent rivets, centerline to centerline, is 4 1/2 times the diameter of one of them.

Thus,

$$1 \text{ space} = 4\frac{1}{2} \times \frac{5}{8}$$

$$= \frac{9}{2} \times \frac{5}{8}$$

$$= \frac{45}{16}$$

There are 4 such spaces between the first and fifth rivets. Therefore, the total distance, D, is found as follows:

$$D = \overset{1}{\cancel{4}} \times \frac{45}{\underset{4}{\cancel{16}}} = \frac{45}{4} = 11\frac{1}{4}$$

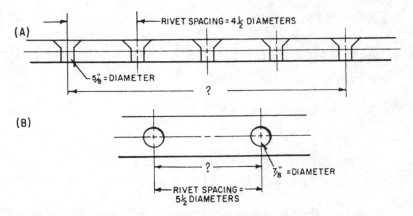

Figure 4-7.—Application of multiplication of fractions in determining rivet spacing.

The distance is $11\dfrac{1}{4}$ inches

Practice problem. Find the distance between the centers of the two rivets shown in figure 4-7 (B).

Answer: $4\dfrac{13}{16}$ inches

DIVISION

There are two methods commonly used for performing division with fractions. One is the common denominator method and the other is the reciprocal method.

Common Denominator Method

The common denominator method is an adaptation of the method of like fractions. The rule is as follows: Change the dividend and divisor to like fractions and divide the numerator of the dividend by the numerator of the divisor. This method can be demonstrated with whole numbers, first changing them to fractions with 1 as the denominator. For example, $12 \div 4$ can be written as follows:

$$12 \div 4 = \frac{12}{1} \div \frac{4}{1}$$

$$= \frac{12 \div 4}{1 \div 1}$$

$$= \frac{12 \div 4}{1}$$

$$= 3$$

If the dividend and divisor are both fractions, as in 1/3 divided by 1/4, we proceed as follows:

$$\frac{1}{3} \div \frac{1}{4} = \frac{4}{12} \div \frac{3}{12}$$

$$= \frac{4 \div 3}{12 \div 12}$$

$$= \frac{4 \div 3}{1}$$

$$= 4 \div 3 = 1\frac{1}{3}$$

Reciprocal Method

The word "reciprocal" denotes an interchangeable relationship. It is used in mathe-

116

matics to describe a specific relationship between two numbers. We say that two numbers are reciprocals of each other if their product is one. In the example $4 \times \frac{1}{4} = 1$, the fractions $\frac{4}{1}$ and $\frac{1}{4}$ are reciprocals. Notice the interchangeability: 4 is the reciprocal of $\frac{1}{4}$ and $\frac{1}{4}$ is the reciprocal of 4.

What is the reciprocal of $\frac{3}{7}$? It must be a number which, when multiplied by $\frac{3}{7}$, produces the product, 1. Therefore,

$$\frac{3}{7} \times \; ? \; = 1$$

$$\frac{\cancel{3}}{\cancel{7}} \times \frac{\cancel{7}}{\cancel{3}} = 1$$

We see that $\frac{7}{3}$ is the only number that could fulfill the requirement. Notice that the numerator and denominator of $\frac{3}{7}$ were simply interchanged to get its reciprocal. If we know a number, we can always find its reciprocal by dividing 1 by the number. Notice this principle in the following examples:

1. What is the reciprocal of 7?

$$1 \div 7 = \frac{1}{7}$$

Check:

$$\frac{\cancel{7}}{\cancel{1}} \times \frac{\cancel{1}}{\cancel{7}} = 1$$

Notice that the cancellation process in this example does not show the usual 1 s which result when dividing a number into itself. For example, when 7 cancels 7, the quotient 1 could be shown beside each of the 7 s. However, since 1 as a factor has the same effect whether it is written in or simply understood, the 1 s need not be written.

2. What is the reciprocal of $\frac{3}{8}$?

$$1 \div \frac{3}{8} = \frac{8}{8} \div \frac{3}{8}$$

$$= 8 \div 3, \text{ or } \frac{8}{3}$$

Check:

$$\frac{\cancel{3}}{\cancel{8}} \times \frac{\cancel{8}}{\cancel{3}} = 1.$$

3. What is the reciprocal of $\frac{5}{2}$?

SOLUTION: $\quad 1 \div \frac{5}{2} = \frac{2}{2} \div \frac{5}{2}$

$$= 2 \div 5$$

$$= \frac{2}{5}$$

Check: $\dfrac{\not5}{\not2} \times \dfrac{\not2}{\not5} = 1$

4. What is the reciprocal of $3\dfrac{1}{8}$?

SOLUTION: $1 \div 3\dfrac{1}{8} = \dfrac{8}{8} \div \dfrac{25}{8}$

$= 8 \div 25$

$= \dfrac{8}{25}$

Check: $\dfrac{2\not5}{\not8} \times \dfrac{\not8}{2\not5} = 1$

The foregoing examples lead to the rule for finding the reciprocal of any number: The reciprocal of a number is the fraction formed when 1 is divided by the number. (If the final result is a whole number, it can be considered as a fraction whose denominator is 1.) A shortcut rule which is purely mechanical and does not involve reasoning may be stated as follows: To find the reciprocal of a number, express the number as a fraction and then invert the fraction.

When the numerator of a fraction is 1, the reciprocal is a whole number. The smaller the fraction, the greater is the reciprocal. For example, the reciprocal of $\dfrac{1}{1,000}$ is 1,000.

Also, the reciprocal of any whole number is a proper fraction. Thus the reciprocal of 50 is

119

$\frac{1}{50}$.

Practice problems. Write the reciprocal of each of the following numbers:

1. 4   2. $\frac{1}{3}$ 3. $2\frac{1}{2}$ 4. 17   5. $\frac{3}{2}$ 6. $\frac{5}{1}$

Answers:

1. $\frac{1}{4}$ 2. 3   3. $\frac{2}{5}$ 4. $\frac{1}{17}$ 5. $\frac{2}{3}$ 6. $\frac{1}{5}$

The reciprocal method of division makes use of the close association of multiplication and division. In any division problem, we must find the answer to the following question: What number multiplied by the divisor yields the dividend? For example, if the problem is to divide 24 by 6, we must find the factor which, when multiplied by 6, yields 24. Experience tells us that the number we seek is 1/6 of 24. Thus, we may rewrite the problem as follows:

$$24 \div 6 = \frac{1}{6} \times 24$$

$$= \frac{1 \times \overset{4}{\cancel{24}}}{\cancel{6} \times 1}$$

$$= 4$$

Check: $6 \times 4 = 24$

In the example $1\frac{1}{2} \div 3$, we could write $3 \times$? =

$1\frac{1}{2}$. The number we seek must be one-third of $1\frac{1}{2}$. Thus we can do the division by taking one-third of $1\frac{1}{2}$; that is, we multiply $1\frac{1}{2}$ by the reciprocal of 3.

$$1\frac{1}{2} \div 3 = 1\frac{1}{2} \times \frac{1}{3}$$

$$= \frac{\cancel{3}}{2} \times \frac{1}{\cancel{3}}$$

$$= \frac{1}{2}$$

Check: $3 \times \frac{1}{2} = \frac{3}{2} = 1\frac{1}{2}$

The rule for division by the reciprocal method is: Multiply the dividend by the reciprocal of the divisor. This is sometimes stated in short form as follows: Invert the divisor and multiply.

The following examples of cases that arise in division with fractions will be solved by both the reciprocal method and the common denominator method. The common denominator method more clearly shows the division process and is easier for the beginner to grasp. The reciprocal method is more obscure as to the reason for its use but has the advantage of speed and the possibility of cancellation of like factors, which simplifies the computation. It is the suggested method once the principles become familiar.

EXAMPLE: $\dfrac{2}{5} \div 4 = ?$

Common Denominator Method

$\dfrac{2}{5} \div 4 = \dfrac{2}{5} \div \dfrac{20}{5}$

$= 2 \div 20$

$= \dfrac{2}{20} = \dfrac{1}{10}$

Reciprocal Method

$\dfrac{2}{5} \div 4 = \dfrac{2}{5} \times \dfrac{1}{4}$

$= \dfrac{\cancel{2} \times 1}{5 \times \cancel{4}}$
$\quad\quad\quad 2$

$= \dfrac{1}{10}$

EXAMPLE: $2\dfrac{2}{3} \div 3 = ?$

Common Denominator Method

$2\dfrac{2}{3} \div 3 = \dfrac{8}{3} \div \dfrac{9}{3}$

$= 8 \div 9$

$= \dfrac{8}{9}$

Reciprocal Method

$2\dfrac{2}{3} \div 3 = \dfrac{8}{3} \times \dfrac{1}{3}$

$= \dfrac{8 \times 1}{3 \times 3}$

$= \dfrac{8}{9}$

EXAMPLE: $9 \div \dfrac{2}{7} = ?$

Common Denominator Method

$$9 \div \frac{2}{7} = \frac{63}{7} \div \frac{2}{7}$$

$$= 63 \div 2$$

$$= \frac{63}{2} = 31\frac{1}{2}$$

Reciprocal Method

$$9 \div \frac{2}{7} = 9 \times \frac{7}{2}$$

$$= \frac{9 \times 7}{1 \times 2}$$

$$= \frac{63}{2} = 31\frac{1}{2}$$

EXAMPLE: $\qquad 10 \div 5\frac{3}{4} = ?$

Common Denominator Method

$$10 \div 5\frac{3}{4} = \frac{40}{4} \div \frac{23}{4}$$

$$= 40 \div 23$$

$$= \frac{40}{23} = 1\frac{17}{23}$$

Reciprocal Method

$$10 \div 5\frac{3}{4} = 10 \times \frac{4}{23}$$

$$= \frac{10 \times 4}{1 \times 23}$$

$$= \frac{40}{23} = 1\frac{17}{23}$$

EXAMPLE: $\qquad \frac{2}{3} \div \frac{1}{4} = ?$

Common Denominator Method

$$\frac{2}{3} \div \frac{1}{4} = \frac{8}{12} \div \frac{3}{12}$$

$$= 8 \div 3$$

Reciprocal Method

$$\frac{2}{3} \div \frac{1}{4} = \frac{2}{3} \times \frac{4}{1}$$

$$= \frac{8}{3} = 2\frac{2}{3}$$

$$= \frac{8}{3} = 2\frac{2}{3}$$

EXAMPLE: $\frac{9}{16} \div \frac{3}{10} = ?$

Common Denominator Method

Reciprocal Method

$$\frac{9}{16} \div \frac{3}{10} = \frac{45}{80} \div \frac{24}{80}$$

$$= 45 \div 24$$

$$= \frac{45}{24} = \frac{15}{8}$$

$$= 1\frac{7}{8}$$

$$\frac{9}{16} \div \frac{3}{10} = \frac{9}{16} \times \frac{10}{3}$$

$$= \frac{\overset{3}{\cancel{9}}}{\underset{8}{\cancel{16}}} \times \frac{\overset{5}{\cancel{10}}}{\cancel{3}}$$

$$= \frac{15}{8} = 1\frac{7}{8}$$

Practice problems. Perform the following division by the reciprocal method:

1. $\frac{3}{8} \div \frac{2}{3}$ 2. $2\frac{1}{3} \div 1\frac{1}{2}$ 3. $\frac{5}{8} \div \frac{5}{16}$ 4. $\frac{1}{3} \div \frac{4}{6}$

Answers:

1. $\frac{9}{16}$ 2. $1\frac{5}{9}$ 3. 2   4. $\frac{1}{2}$

COMPLEX FRACTIONS

When the numerator or denominator, or both, in a fraction are themselves composed of fractions, the resulting expression is called a complex fraction. The following expression is a complex fraction:

124

$$\frac{3/5}{3/4}$$

This should be read "three-fifths over three-fourths" or "three-fifths divided by three-fourths." Any complex fraction may be simplified by writing it as a division problem, as follows:

$$\frac{3/5}{3/4} = \frac{3}{5} \div \frac{3}{4}$$

$$= \frac{\cancel{3}}{5} \cdot \frac{4}{\cancel{3}}$$

$$= 4/5$$

Similarly,

$$\frac{3\frac{1}{3}}{2\frac{1}{2}} = \frac{10}{3} \div \frac{5}{2} = \frac{\cancel{10}^{2}}{3} \times \frac{2}{\cancel{5}} = \frac{4}{3} = 1\frac{1}{3}$$

Complex fractions may also contain an indicated operation in the numerator or denominator or both. Thus,

$$\frac{\frac{1}{2} + \frac{1}{3}}{\frac{9}{5} + \frac{1}{5}}$$

is a complex fraction. To simplify such a fraction we simplify the numerator and denominator and proceed as follows:

$$\frac{\dfrac{1}{2}+\dfrac{1}{3}}{\dfrac{9}{5}+\dfrac{1}{5}} = \frac{\dfrac{3}{6}+\dfrac{2}{6}}{\dfrac{10}{5}} = \frac{\dfrac{5}{6}}{2}$$

$$= \frac{5}{6} \div \frac{2}{1}$$

$$= \frac{5}{6} \times \frac{1}{2}$$

$$= \frac{5}{12}$$

Mixed numbers appearing in complex fractions usually show the plus sign.

Thus,

$$4\frac{2}{5} \div 7\frac{1}{3}$$

might be written

$$\frac{4+\dfrac{2}{5}}{7+\dfrac{1}{3}}$$

Practice problems. Simplify the following complex fractions:

1. $\dfrac{\dfrac{1}{3}}{\dfrac{3}{8}}$ 2. $\dfrac{2\dfrac{1}{2}}{3}$ 3. $\dfrac{3\dfrac{2}{3}}{2\dfrac{2}{5}}$ 4. $\dfrac{\dfrac{1}{4}+\dfrac{1}{3}}{\dfrac{1}{16}-\dfrac{1}{32}}$

Answers:

1. $\frac{8}{3}$ 2. $\frac{5}{6}$ 3. $1\frac{19}{36}$ 4. $18\frac{2}{3}$

Complex fractions may arise in electronics when it is necessary to find the total resistance of several resistances in parallel as shown in figure 4-8. The rule is: The total resistance of a parallel circuit is 1 divided by the sum of the reciprocals of the separate resistances. Written as a formula, this produces the following expression:

$$R_t = \frac{1}{\frac{1}{R_1} + \frac{1}{R_2} + \frac{1}{R_3}}$$

EXAMPLE: Find the total resistance of the parallel circuit in figure 4-8 (A). Substituting the values 3, 4, and 6 for the letters R_1, R_2, and R_3, we have the following:

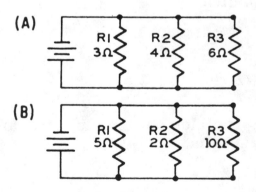

Figure 4-8.—Application of complex fractions in calculating electrical resistance.

$$R_t = \cfrac{1}{\frac{1}{3} + \frac{1}{4} + \frac{1}{6}}$$

The LCD of the fractions $\frac{1}{3}$, $\frac{1}{4}$, and $\frac{1}{6}$ is 12.

Thus,

$$R_t = \cfrac{1}{\frac{4}{12} + \frac{3}{12} + \frac{2}{12}}$$

$$= \cfrac{1}{\frac{9}{12}}$$

$$= \frac{12}{9} = \frac{4}{3}$$

$$= 1\frac{1}{3} \text{ ohms (measure of resistance).}$$

Practice problem: Find the total resistance of the parallel circuit in figure 4-8 (B).

Answer: $1\frac{1}{4}$ ohms.

CHAPTER 5

DECIMALS

The origin and meaning of the word "decimal" were discussed in chapter 1 of this course. Also discussed in chapter 1 were the concept of place value and the use of the number ten as the base for our number system. Another term which is frequently used to denote the base of a number system is RADIX. For example, two is the radix of the binary system and ten is the radix of the decimal system. The radix of a number system is always equal to the number of different digits used in the system. For example, the decimal system, with radix ten, has ten digits: 0 through 9.

DECIMAL FRACTIONS

A decimal fraction is a fraction whose denominator is 10 or some power of 10, such as 100, 1,000, or 10,000. Thus, $\frac{7}{10}$, $\frac{12}{100}$, and $\frac{215}{1000}$ are decimal fractions. Decimal fractions have special characteristics that make computation much simpler than with other fractions.

Decimal fractions complete our decimal

system of numbers. In the study of whole numbers, we found that we could proceed to the left from the units place, tens, hundreds, thousands, and on indefinitely to any larger place value, but the development stopped with the units place. Decimal fractions complete the development so that we can proceed to the right of the units place to any smaller number indefinitely.

Figure 5-1 (A) shows how decimal fractions complete the system. It should be noted that as we proceed from left to right, the value of each place is one-tenth the value of the preceding place, and that the system continues uninterrupted with the decimal fractions.

Figure 5-1 (B) shows the system again, this time using numbers. Notice in (A) and (B) that the units place is the center of the system and that the place values proceed to the right or left of it by powers of ten. Ten on the left is balanced by tenths on the right, hundreds by hundredths, thousands by thousandths, etc.

Notice that 1/10 is one place to the right of the units digit, 1/100 is two places to the right, etc. (See fig. 5-1.) If a marker is placed after the units digit, we can decide whether a decimal digit is in the tenths, hundredths, or thousandths position by counting places to the right of the marker. In some European countries, the marker is a comma; but in the English-speaking countries, the marker is the DECIMAL POINT.

Thus, $\frac{3}{10}$ is written 0.3. To write $\frac{3}{100}$ it is necessary to show that 3 is in the second place to the right of the decimal point, so a zero is inserted in the first place. Thus, $\frac{3}{100}$ is written

0.03. Similarly, $\frac{3}{1000}$ can be written by insert-

ing zeros in the first two places to the right of

the decimal point. Thus, $\frac{3}{1000}$ is written 0.003.

In the number 0.3, we say that 3 is in the first
decimal place; in 0.03, 3 is in the second decimal place; and in 0.003, 3 is in the third decimal place. Quiet frequently decimal fractions
are simply called decimals when written in this
shortened form.

WRITING DECIMALS

Any decimal fraction may be written in the
shortened form by a simple mechanical process.
Simply begin at the right-hand digit of the numerator and count off to the left as many places
as there are zeros in the denominator. Place
the decimal point to the left of the last digit
counted. The denominator may then be disregarded. If there are not enough digits, as
many place-holding zeros as are necessary are
added to the left of the left-hand digit in the
numerator.

Thus, in $\frac{23}{10000}$, beginning with the digit 3,

we count off four places to the left, adding two
0's as we count, and place the decimal point to
the extreme left. (See fig. 5-2.) Either form
is read "twenty-three ten-thousandths."

When a decimal fraction is written in the
shortened form, there will always be as many
decimal places in the shortened form as there

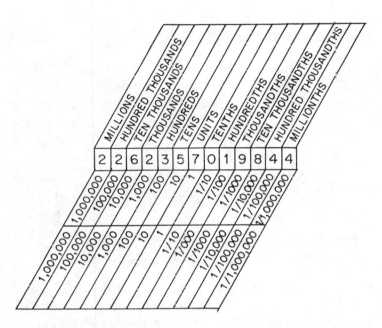

Figure 5-1.—Place values including decimals.

$$\frac{23}{10000} = \overset{4 \leftarrow 3 \leftarrow 2 \leftarrow 1}{.0023}$$

PLACE HOLDING
ZEROS ADDED

Figure 5-2.—Conversion
of a decimal fraction
to shortened form.

are zeros in the denominator of the fractional form.

Figure 5-3 shows the fraction $\frac{24358}{100000}$ and what is meant when it is changed to the shortened form. This figure is presented to show further that each digit of a decimal fraction holds a certain position in the digit sequence and has a particular value.

By the fundamental rule of fractions, it

132

should be clear that $\dfrac{5}{10} = \dfrac{50}{100} = \dfrac{500}{1000}$. Writing the same values in the shortened way, we have 0.5 = 0.50 = 0.500. In other words, the value of a decimal is not changed by annexing zeros at the right-hand end of the number. This is not

$$\dfrac{24{,}358}{100{,}000} \quad \text{ALSO MEANS} \atop \text{THE SUM OF} \left\{ \begin{array}{lll} \text{2 TENTHS} & \text{OR .2} \\ \text{4 HUNDREDTHS} & \text{OR .04} \\ \text{3 THOUSANDTHS} & \text{OR .003} \\ \text{5 TEN-THOUSANDTHS} & \text{OR .0005} \\ \text{8 HUNDRED-THOUSANDTHS} & \text{OR .00008} \end{array} \right.$$

.24358

Figure 5-3.—Steps in the conversion of a
decimal fraction to shortened form.

true of whole numbers. Thus, 0.3, 0.30, and 0.300 are equal but 3, 30, and 300 are not equal. Also notice that zeros directly after the decimal point do change values. Thus 0.3 is not equal to either 0.03 or 0.003.

Decimals such as 0.125 are frequently seen. Although the 0 on the left of the decimal point is not required, it is often helpful. This is particularly true in an expression such as 32 ÷ 0.1. In this expression, the lower dot of the division symbol must not be crowded against the decimal point; the 0 serves as an effective spacer. If any doubt exists concerning the clarity of an expression such as .125, it should be written as 0.125.

Practice problems. In problems 1 through 4, change the fractions to decimals. In problems 5 through 8, write the given numbers as decimals:

1. 8/100

2. 5/1000

3. 43/1000
4. 32/10000
5. Four hundredths
6. Four thousandths
7. Five hundred one ten-thousandths
8. Ninety-seven thousandths

Answers:

1. 0.08	5. 0.04
2. 0.005	6. 0.004
3. 0.043	7. 0.0501
4. 0.0032	8. 0.097

READING DECIMALS

To read a decimal fraction in full, we read both its numerator and denominator, as in reading common fractions. To read 0.305, we read "three hundred five thousandths." The denominator is always 1 with as many zeros as decimal places. Thus the denominator for 0.14 is 1 with two zeros, or 100. For 0.003 it is 1,000; for 0.101 it is 1,000; and for 0.3 it is 10. The denominator may also be determined by counting off place values of the decimal. For 0.13 we may think "tenths, hundredths" and the fraction is in hundredths. In the example 0.1276 we may think "tenths, hundredths, thousandths, ten-thousandths." We see that the denominator is 10,000 and we read the fraction "one thousand two hundred seventy-six ten-thousandths."

A whole number with a fraction in the form of a decimal is called a MIXED DECIMAL.

Mixed decimals are read in the same manner as mixed numbers. We read the whole number in the usual way followed by the word "and" and then read the decimal. Thus, 160.32 is read "one hundred sixty and thirty-two hundredths." The word "and" in this case, as with mixed numbers, means plus. The number 3.2 means three plus two tenths.

It is also possible to have a complex decimal. A COMPLEX DECIMAL contains a common fraction. The number $0.3\frac{1}{3}$ is a complex decimal and is read "three and one-third tenths." The number $0.87\frac{1}{2}$ means $87\frac{1}{2}$ hundredths. The common fraction in each case forms a part of the last or right-hand place.

In actual practice when numbers are called out for recording, the above procedure is not used. Instead, the digits are merely called out in order with the proper placing of the decimal point. For example, the number 216.003 is read, "two one six point zero zero three." The number 0.05 is read, "zero point zero five."

EQUIVALENT DECIMALS

Decimal fractions may be changed to equivalent fractions of higher or lower terms, as is the case with common fractions. If each decimal fraction is rewritten in its common fraction form, changing to higher terms is accomplished by multiplying both numerator and denominator by 10, or 100, or some higher power of 10. For example, if we desire to change $\frac{5}{10}$ to hundredths, we may do so by multiplying both numerator and denominator by 10. Thus,

$$\frac{5}{10} = \frac{50}{100}$$

In the decimal form, the same thing may be accomplished by simply annexing a zero. Thus,

$$0.5 = 0.50$$

Annexing a 0 on a decimal has the same effect as multiplying the common fraction form of the decimal by 10/10. This is an application of the fundamental rule of fractions. Annexing two 0's has the same effect as multiplying the common fraction form of the decimal by 100/100; annexing three 0 s has the same effect as multiplying by 1000/1000; etc.

REDUCTION TO LOWER TERMS

Reducing to lower terms is known as ROUND-OFF, or simply ROUNDING, when dealing with decimal fractions. If it is desired to reduce 6.3000 to lower terms, we may simply drop as many end zeros as necessary since this is equivalent to dividing both terms of the fraction by some power of ten. Thus, we see that 6.3000 is the same as 6.300, 6.30, or 6.3.

It is frequently necessary to reduce a number such as 6.427 to some lesser degree of precision. For example, suppose that 6.427 is to be rounded to the nearest hundredth. The question to be decided is whether 6.427 is closer to 6.42 or 6.43. The best way to decide this question is to compare the fractions 420/1000, 427/1000, and 430/1000. It is obvious that 427/1000 is closer to 430/1000, and 430/1000 is equivalent to 43/100; therefore we say that 6.427, correct to the nearest hundredth, is 6.43.

A mechanical rule for rounding off can be developed from the foregoing analysis. Since the digit in the tenths place is not affected when we round 6.427 to hundredths, we may limit our attention to the digits in the hundredths and thousandths places. Thus the decision reduces to the question whether 27 is closer to 20 or 30. Noting that 25 is halfway between 20 and 30, it is clear that anything greater than 25 is closer to 30 than it is to 20.

In any number between 20 and 30, if the digit in the thousandths place is greater than 5, then the number formed by the hundredths and thousandths digits is greater than 25. Thus we would round the 27 in our original problem to 30, as far as the hundredths and thousandths digits are concerned. This result could be summarized as follows: When rounding to hundredths, if the digit in the thousandths place is greater than 5, increase the digit in the hundredths place by 1 and drop the digit in the thousandths place.

The digit in the thousandths place may be any one of the ten digits, 0 through 9. If these ten digits are split into two groups, one composed of the five smaller digits (0 through 4) and the other composed of the five larger digits, then 5 is counted as one of the larger digits. Therefore, the general rule for rounding off is stated as follows: If the digit in the decimal place to be eliminated is 5 or greater, increase the digit in the next decimal place to the left by 1. If the digit to be eliminated is less than 5, leave the retained digits unchanged.

The following examples illustrate the rule for rounding off:

1. 0.1414 rounded to thousandths is 0.141.
2. 3.147 rounded to tenths is 3.1.
3. 475 rounded to the nearest hundred is 500.

Observe carefully that the answer to example 2 is not 3.2. Some trainees make the error of treating the rounding process as a kind of chain reaction, in which one first rounds 3.147 to 3.15 and then rounds 3.15 to 3.2. The error of this method is apparent when we note that 147/1000 is closer to 100/1000 than it is to 200/1000.

Problems of the following type are sometimes confusing: Reduce 2.998 to the nearest hundredth. To drop the end figure we must increase the next figure by 1. The final result is 3.00. We retain the zeros to show that the answer is carried to the nearest hundredth.

Practice problems. Round off as indicated:

1. 0.5862 to hundredths
2. 0.345 to tenths
3. 2346 to hundreds
4. 3.999 to hundredths
   Answers:

1. 0.59
2. 0.3

3. 2300
4. 4.00

CHANGING DECIMALS
TO COMMON FRACTIONS

Any decimal may be reduced to a common fraction. To do this we simply write out the numerator and denominator in full and reduce to lowest terms. For example, to change 0.12 to a common fraction, we simply write out the fraction in full,

$$\frac{12}{100}$$

and reduce to lowest terms,

$$\frac{\overset{3}{\cancel{12}}}{\underset{25}{\cancel{100}}} = \frac{3}{25}$$

Likewise, 0.77 is written

$$\frac{77}{100}$$

but this is in lowest terms so the fraction cannot be further reduced.

One way of checking to see if a decimal fraction can be reduced to lower terms is to consider the makeup of the decimal denominator. The denominator is always 10 or a power of 10. Inspection shows that the prime factors of 10 are 5 and 2. Thus, the numerator must be divisible by 5 or 2 or both, or the fraction cannot be reduced.

EXAMPLE: Change the decimal 0.0625 to a common fraction and reduce to lowest terms.

SOLUTION: $0.0625 = \dfrac{625}{10000}$

$$= \dfrac{625 \div 25}{10000 \div 25} = \dfrac{25}{400}$$

$$= \dfrac{1}{16}$$

Complex decimals are changed to common fractions by first writing out the numerator and denominator in full and then reducing the resulting complex fraction in the usual way. For example, to reduce $0.12\frac{1}{2}$, we first write

$$\dfrac{12\frac{1}{2}}{100}$$

Writing the numerator as an improper fraction we have

$$\dfrac{\frac{25}{2}}{100}$$

and applying the reciprocal method of division, we have

$$\dfrac{25}{2} \times \dfrac{1}{\cancel{100}\,4} = \dfrac{1}{8}$$

140

Practice problems. Change the following decimals to common fractions in lowest terms:

1. 0.25

3. $0.6\frac{1}{4}$

2. 0.375

4. $0.03\frac{1}{5}$

Answers:

1. 1/4   3. 5/8
2. 3/8   4. 4/125

CHANGING COMMON
FRACTIONS TO DECIMALS

The only difference between a decimal fraction and a common fraction is that the decimal fraction has 1 with a certain number of zeros (in other words, a power of 10) for a denominator. Thus, a common fraction can be changed to a decimal if it can be reduced to a fraction having a power of 10 for a denominator.

If the denominator of the common fraction in its lowest terms is made up of the prime factors 2 or 5 or both, the fraction can be converted to an exact decimal. If some other prime factor is present, the fraction cannot be converted exactly. The truth of this is evident when we consider the denominator of the new fraction. It must always be 10 or a power of 10, and we know the factors of such a number are always 2 s and 5 s.

The method of converting a common fraction

to a decimal is illustrated as follows:

EXAMPLE: Convert 3/4 to a decimal.

SOLUTION:
$$\frac{3}{4} = \frac{300}{400}$$

$$= \frac{300}{4} \times \frac{1}{100}$$

$$= 75 \times \frac{1}{100}$$

$$= 0.75$$

Notice that the original fraction could have been rewritten as 3000/4000, in which case the result would have been 0.750. On the other hand, if the original fraction had been rewritten as 30/40, the resulting division of 4 into 30 would not have been possible without a remainder. When the denominator in the original fraction has only 2 s and 5 s as factors, so that we know a remainder is not necessary, the fraction should be rewritten with enough 0 s to complete the division with no remainder.

Observation of the results in the foregoing example leads to a shortcut in the conversion method. Noting that the factor 1/100 ultimately enters the answer in the form of a decimal, we could introduce the decimal point as the final step without ever writing the fraction 1/100. Thus the rule for changing fractions to decimals is as follows:

1. Annex enough 0 s to the numerator of the original fraction so that the division will be

exact (no remainder).

2. Divide the original denominator into the new numerator formed by annexing the 0 s.

3. Place the decimal point in the answer so that the number of decimal places in the answer is the same as the number of 0 s annexed to the original numerator.

If a mixed number in common fraction form is to be converted, convert only the fractional part and then write the two parts together. This is illustrated as follows:

$$2\frac{3}{4} = 2 + \frac{3}{4} = 2 + .75 = 2.75$$

Practice problems. Convert the following common fractions and mixed numbers to decimal form:

1. $\frac{1}{4}$ 2. $\frac{3}{8}$ 3. $\frac{5}{32}$ 4. $2\frac{5}{16}$

Answers:

1. 0.25   2. 0.375   3. 0.15625   4. 2.3125

Nonterminating Decimals

As stated previously, if the denominator of a common fraction contains some prime factor other than 2 or 5, the fraction cannot be converted completely to a decimal. When such fractions are converted according to the foregoing rule, the decimal resulting will never terminate. Consider the fraction 1/3. Applying the rule, we have

143

$$
\begin{array}{r}
.333 \ \ldots \\
3\overline{)1.0000} \\
\underline{9} \\
10 \\
\underline{9} \\
10 \\
\underline{9}
\end{array}
$$

The division will continue indefinitely. Any common fraction that cannot be converted exactly yields a decimal that will never terminate and in which the digits sooner or later recur. In the previous example, the recurring digit was 3. In the fraction 5/11, we have

$$
\begin{array}{r}
.4545 \\
11\overline{)5.0000} \\
\underline{4\,4} \\
60 \\
\underline{55} \\
50 \\
\underline{44} \\
60 \\
\underline{55}
\end{array}
$$

The recurring digits are 4 and 5.

When a common fraction generates such a repeating decimal, it becomes necessary to arbitrarily select a point at which to cease the repetition. This may be done in two ways. We may write the decimal fraction by rounding off at the desired point. For example, to round off the decimal generated by $\frac{1}{3}$ to hundredths, we carry the division to thousandths, see that this figure is less than 5, and drop it. Thus, $\frac{1}{3}$ rounded to hundredths is 0.33. The other method is to carry the division to the desired number of decimal places and carry the remaining incomplete division as a common fraction—that is, we write the result of a complex decimal. For example, $\frac{1}{3}$ carried to thousandths would be

$$\frac{1}{3} = 3\overline{)1.000}^{\textstyle .333\frac{1}{3}}$$
$$\begin{array}{r} \underline{9} \\ 10 \\ \underline{9} \\ 10 \\ \underline{9} \\ 1 \end{array}$$

Practice problems. Change the following common fractions to decimals with three places and carry the incomplete division as a common fraction:

1. $\frac{7}{13}$ 2. $\frac{5}{9}$ 3. $\frac{4}{15}$ 4. $\frac{5}{12}$

Answers:

1. $0.538\frac{6}{13}$ 3. $0.266\frac{2}{3}$

2. $0.555\frac{5}{9}$ 4. $0.416\frac{2}{3}$

OPERATION WITH DECIMALS

In the study of addition of whole numbers, it was established that units must be added to units, tens to tens, hundreds to hundreds, etc. For convenience, in adding several numbers, units were written under units, tens under tens, etc. The addition of decimals is accomplished in the same manner.

ADDITION

In adding decimals, tenths are written under tenths, hundredths under hundredths, etc. When this is done, the decimal points fall in a straight line. The addition is the same as in adding whole numbers. Consider the following example:

$$\begin{array}{r} 2.18 \\ 34.35 \\ 0.14 \\ 4.90 \\ \hline 41.57 \end{array}$$

Adding the first column on the right gives 17 hundredths or 1 tenth and 7 hundredths. As with whole numbers, we write the 7 under the hundredths column and add the 1 tenth in the tenths column—that is, the column of the next higher order. The sum of the tenths column is 15 tenths or 1 unit and 5 tenths. The 5 is written under the tenths column and the 1 is added in the units column.

It is evident that if the decimal points are kept in a straight line—that is, if the place values are kept in the proper columns—addition with decimals may be accomplished in the ordinary manner of addition of whole numbers. It should also be noted that the decimal point of the sum falls directly under the decimal points of the addends.

SUBTRACTION

Subtraction of decimals likewise involves no new principles. Notice that the place values of the subtrahend in the following example are fixed directly under the corresponding place values in the minuend. Notice also that this causes the decimal points to be alined and that the figures in the difference (answer) also retain the correct columnar alinement.

$$\begin{array}{r} 45.76 \\ -31.87 \\ \hline 13.89 \end{array}$$

We subtract column by column, as with whole numbers, beginning at the right.

Practice problems. Add or subtract as indicated:

1. 12.3 + 2.13 + 4 + 1.234
2. 0.5 + 0.04 + 12.001 + 10
3. 237.5 - 217.9
4. 9.04 - 7.156

Answers:

1. 19.664   3. 19.6
2. 22.541   4. 1.884

MULTIPLICATION

Multiplication of a decimal by a whole number may be explained by expressing the decimal as a fraction.

EXAMPLE: Multiply 6.12 by 4.

SOLUTION: $\frac{4}{1} \times \frac{612}{100} = \frac{2448}{100}$

$$= 24.48$$

When we perform the multiplication keeping the decimal form, we have

$$\begin{array}{r} 6.12 \\ 4 \\ \hline 24.48 \end{array}$$

By common sense, it is apparent that the whole number 4 times the whole number 6, with some fraction, will yield a number in the neighborhood of 24. Hence, the placing of the decimal point is reasonable.

An examination of several examples will reveal that the product of a decimal and a whole number has the same number of decimal places as the factor containing the decimal. Zeros, if any, at the end of the decimal should be rejected.

Multiplication of Two Decimals

To show the rule for multiplying two decimals together, we multiply the decimal in fractional form first and then in the conventional

148

way, as in the following example:

$$0.4 \times 0.37$$

Writing these decimals as common fractions, we have

$$\frac{4}{10} \times \frac{37}{100} = \frac{4 \times 37}{10 \times 100}$$

$$= \frac{148}{1000}$$

$$= 0.148$$

In decimal form the problem is

$$\begin{array}{r} 0.37 \\ 0.4 \\ \hline 0.148 \end{array}$$

The placing of the decimal point is reasonable, since 4 tenths of 37 hundredths is a little less than half of 37 hundredths, or about 15 hundredths.

Consider the following example:

$$4.316 \times 3.4$$

In the common fraction form, we have

$$\frac{4316}{1000} \times \frac{34}{10} = \frac{4316 \times 34}{1000 \times 10}$$

$$= \frac{146744}{10000}$$

$$= 14.6744$$

In the decimal form the problem is

$$
\begin{array}{r}
4.316 \\
\underline{3.4} \\
17264 \\
12948 \\
\hline
14.6744
\end{array}
$$

We note that 4 and a fraction times 3 and a fraction yields a product in the neighborhood of 12. Thus, the decimal point is in the logical place.

In the above examples it should be noted in each case that when we multiply the decimals together we are multiplying the numerators. When we place the decimal point by adding the number of decimal places in the multiplier and multiplicand, we are in effect multiplying the denominators.

When the numbers multiplied together are thought of as the numerators, the decimal points may be temporarily disregarded and the numbers may be considered whole. This justifies the apparent disregard for place value in the multiplication of decimals. We see that the rule for multiplying decimals is only a modification of the rule for multiplying fractions.

To multiply numbers in which one or more of the factors contain a decimal, multiply as though the numbers were whole numbers. Mark off as many decimal places in the product as there are decimal places in the factors together.

Practice problems. Multiply as indicated:

1. 3.7 x 0.02   2. 0.45 x 0.7

3.   6.5   4. 0.0073
    x0.01   x5.4
    ―――――   ―――――

Answers:

1. 0.074   2. 0.315

3. 0.065   4. 0.03942

Multiplying by Powers of 10

Multiplying by a power of 10 (10, 100, 1,000, etc.) is done mechanically by simply moving the decimal point to the right as many places as there are zeros in the multiplier. For example, 0.00687 is multiplied by 1,000 by moving the decimal point three places to the right as follows:

$$1,000 \times 0.00687 = 6.87$$

Multiplying a number by 0.1, 0.01, 0.001, etc., is done mechanically by simply moving the decimal point to the left as many places as there are decimal places in the multiplier. For example, 348.2 is multiplied by 0.001 by moving the decimal point three places to the left as follows:

$$348.2 \times 0.001 = 0.3482$$

DIVISION

When the dividend is a whole number, we recognize the problem of division as that of converting a common fraction to a decimal. Thus in the example 5 ÷ 8, we recall that the

151

problem could be written

$$\frac{5000}{1000} \div 8 = \frac{5000 \div 8}{1000}$$

$$= \frac{625}{1000}$$

$$= .625$$

This same problem may be worked by the following, more direct method:

$$\frac{5}{8} = 8\overline{)5.000} \quad \begin{array}{r} .625 \\ \hline 4\ 8 \\ \hline 20 \\ 16 \\ \hline 40 \\ 40 \\ \hline \end{array}$$

Since not all decimals generated by division terminate so early as that in the above example, if at all, it should be predetermined as to how many decimal places it is desired to carry the quotient. If it is decided to terminate a quotient at the third decimal place, the division should be carried to the fourth place so that the correct rounding off to the third place may be determined.

When the dividend contains a decimal, the same procedure applies as when the dividend is whole. Notice the following examples (rounded to three decimal places):

152

1. $6.31 \div 8$

$$
\begin{array}{r}
.7887 = .789 \\
8\overline{\smash{)}6.3100} \\
\underline{5\ 6} \\
71 \\
\underline{64} \\
70 \\
\underline{64} \\
60 \\
\underline{56} \\
4
\end{array}
$$

2. $0.0288 \div 32$

$$
\begin{array}{r}
0.0009 = 0.001 \\
32\overline{\smash{)}0.0288} \\
\underline{288}
\end{array}
$$

Observe in each case (including the case where the dividend is whole), that the quotient contains the same number of decimal places as the number used in the dividend. Notice also that the place values are rigid; that is, tenths in the quotient appear over tenths in the dividend, hundredths over hundredths, etc.

Practice problems. In the following division problems, round off each quotient correct to three decimal places.

1. $10 \div 6$ 3. $2.743 \div 77$

2. $23.5 \div 16$ 4. $1.00 \div 3$

Answers:

1. 1.667   3. 0.036

2. 1.469   4. 0.333

153